SCHOONERS

Robert Shipley/Fred Addis

Great Lakes Album Series

Vanwell Publishing Limited
St. Catharines, Ontario

This book is dedicated to all those
who sailed the schooners of the Great Lakes.

Canadian Cataloguing in Publication Data

Shipley, Robert, 1948-
Schooners

(Great Lakes album series)
1st Canadian ed.
ISBN 0-920277-59-4

1. Schooners - Great Lakes - History - Pictorial works. I. Addis, Fred A. (Fred Arthur), 1952- II. Title. III. Series.

VM311.F7S48 1990 387.2'24'0977 C90-094890-6

Design: Susan Nicholson

Cover: Capt. Richard Goldring's (see "G" burgee on foremast) two masted schooner *Maple Leaf* featured a spoon bow in this C.I. Gibbon painting prior to 1885. Following a fire at Toronto the schooner was rebuilt with a clipper bow and survived until 1926. *Casselman Collection*

CONTENTS

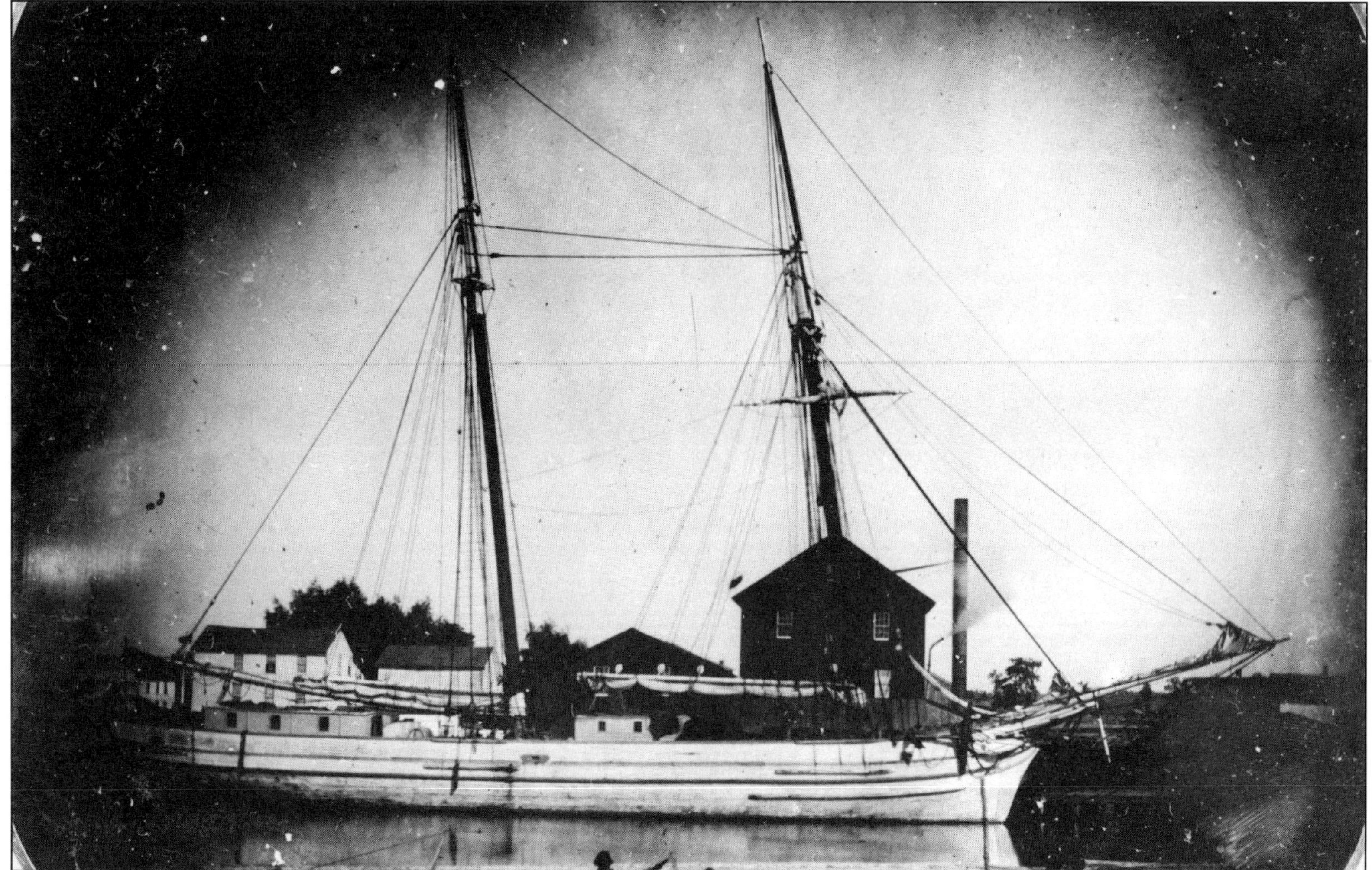

Britannia

The French, British and Americans had built sailing vessels on the Great Lakes in the eighteenth and early nineteenth centuries, but these had either been for communication between isolated outposts or for military service. Only after the War of 1812 did truly commercial shipping begin. The Britannia, built in 1821 by Amos S. Roberts at the west end of Lake Ontario, was one of the first such ships. Its masts were raked or slanted slightly aft. This rig was similar to contemporary oceangoing schooners built on the east coast of North America.

SCHOONERS ON THE GREAT LAKES

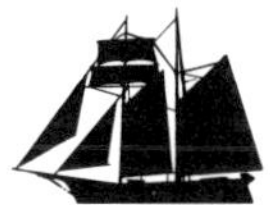

A schooner can be described as a sailing vessel rigged with fore and aft sails on its two or more masts. Fore and aft refers to the sails' normal position along the centreline of the ship as opposed to square-rigged vessels on which the sails were usually across the centreline. First prized by navies as fast dispatch vessels, schooners gained a foothold in the coastal trades and fisheries of early Canada and America. Schooners could be handled by small crews and thus less expensively than their square-rigged counterparts, while carrying the same amount of cargo. The schooner's seakeeping qualities were favoured for its ability to sail close to the wind, that is, the ability to move almost directly (within 15 to 20 degrees) into the direction of the wind.

Schooners were important in the settlement of the Great Lakes region. Initially they carried passengers and finished goods west. By the 1850s railways and the more reliable steamboats had captured those markets, but schooners were still used in the movement of bulk commodities such as lumber, coal and grain.

Because of their suitability for moving bulk cargoes, the innovations and adaptations applied to oceangoing schooners in the interest of speed and seakeeping were not always adopted on the lakes. The limiting size of the canal locks required that lake schooners be blunt in the bow and virtually flat-bottomed. Schooner builders such as Burger and Bates of Manitowoc, Wisconsin, are credited with building the fastest and most visually pleasing schooners. Shipbuilders Shickluna and Muir, located on the Welland Canal, built more functional, if somewhat less attractive, craft.

Before 1899, over three thousand schooners were in active service on the Great Lakes, whether sailing or as barges. As iron and steel shipbuilding took hold in the region and the demand for larger vessels increased, the wooden schooner's days were numbered.

Schooner owners had to be increasingly creative to find work for their ships. Maintenance was expensive and insurance rates soared. Even the insurance they did carry was usually cancelled around the first of November when the lakes' storm season threatened safe navigation. By 1930 fewer than six schooners were still working on the Great Lakes. Their importance eclipsed by the steamboat, the days of wooden ships and iron men and women had come to a close.

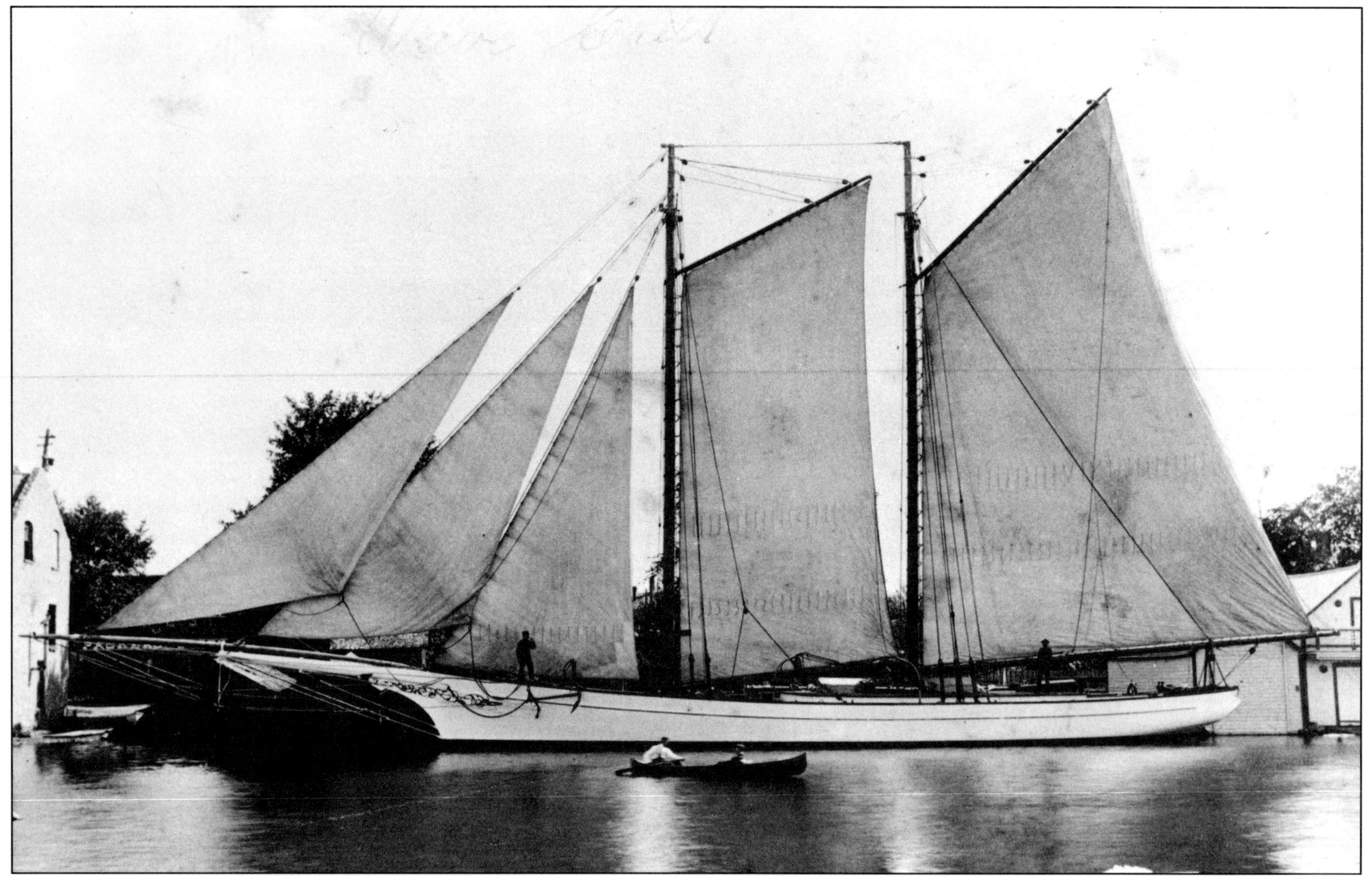

Wave Crest and *Island Queen*

Marine Museum of Upper Canada, Toronto

The *Wave Crest,* shown at Belleville, Ontario, and *Island Queen* in the sunset probably on the Detroit River, look more like yachts than the working boats they were. They represent the purest form of two-masted schooner. *Island Queen* is towing a flat scow that must have been related to her employment at the time the picture was taken. *Wave Crest* worked in the carrying trade around the Bay of Quinte in eastern Lake Ontario.

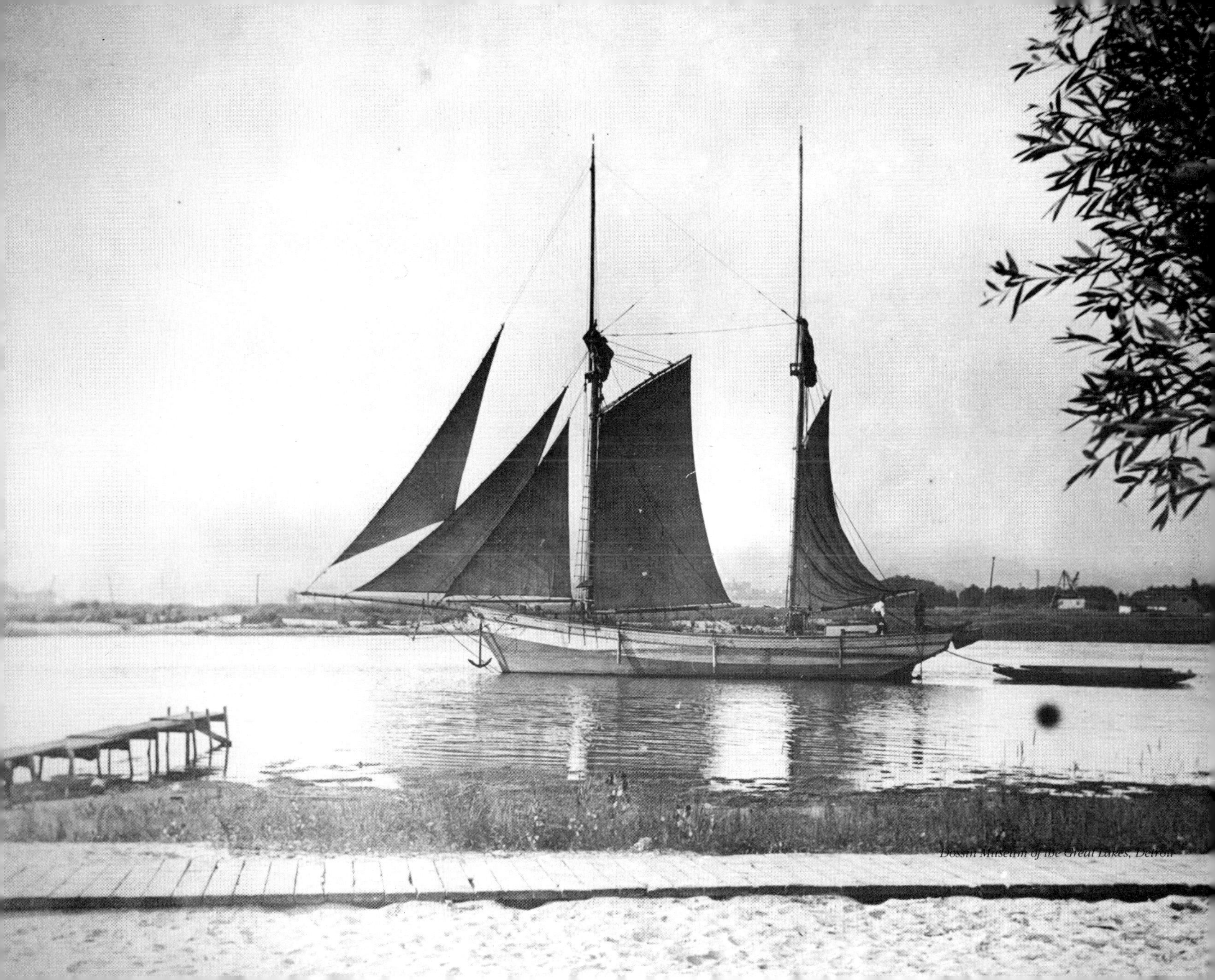

Dossin Museum of the Great Lakes, Detroit

Edward E. Skeele

By the 1850s, when the *Pauline* was launched at the Barber Shipyard in Milwaukee, Wisconsin, the Great Lakes three-masted schooner had reached a fairly advanced degree of development. The long bowsprit and relatively tall masts are evident, and the hull was well suited to carrying its principle cargo, lumber. The vessel was already older than timber magnate Edward Ellms Skeele when he purchased and renamed it in 1914. It continued to work into the 1920s.

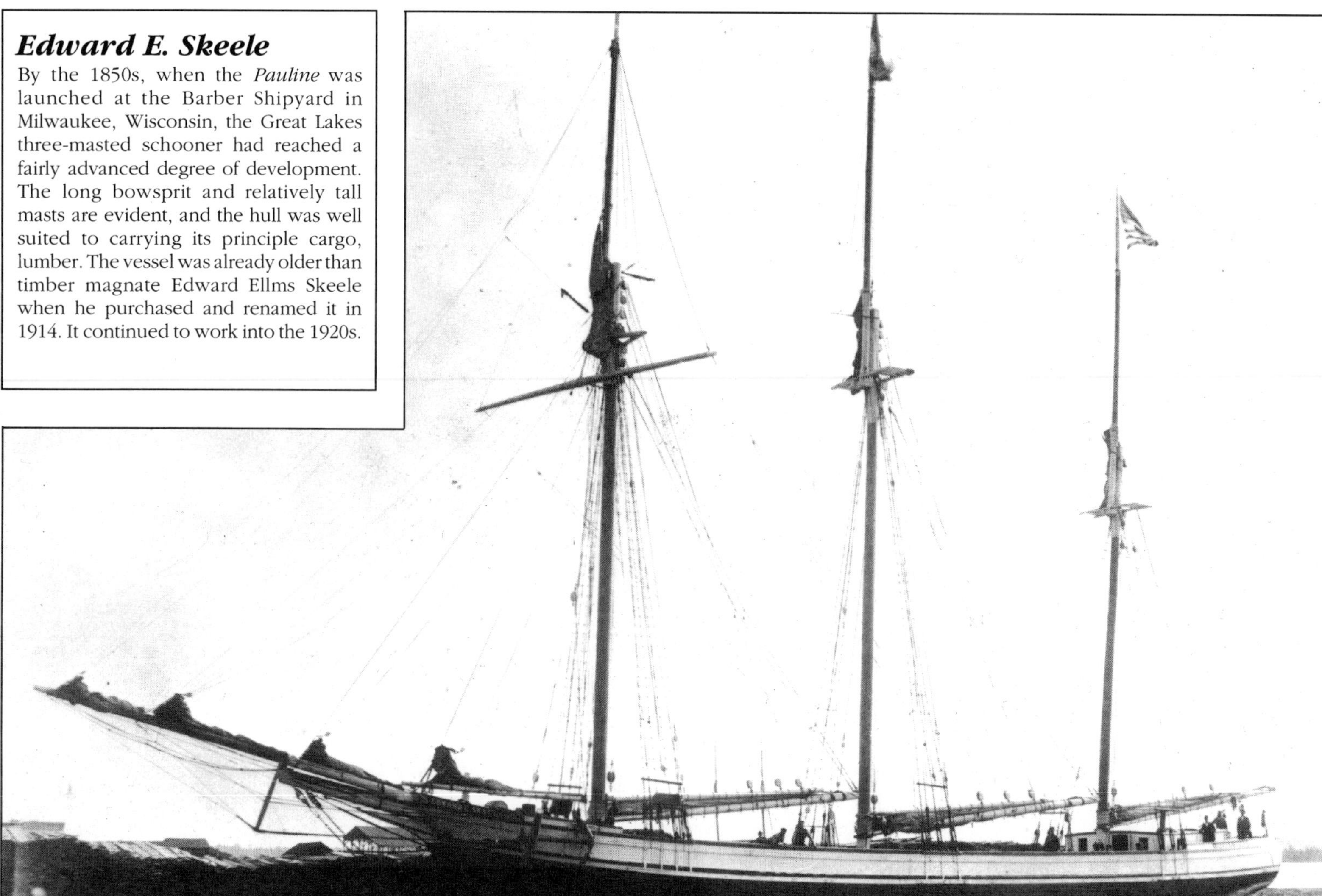

Institute for Great Lakes Research, Bowling Green State University

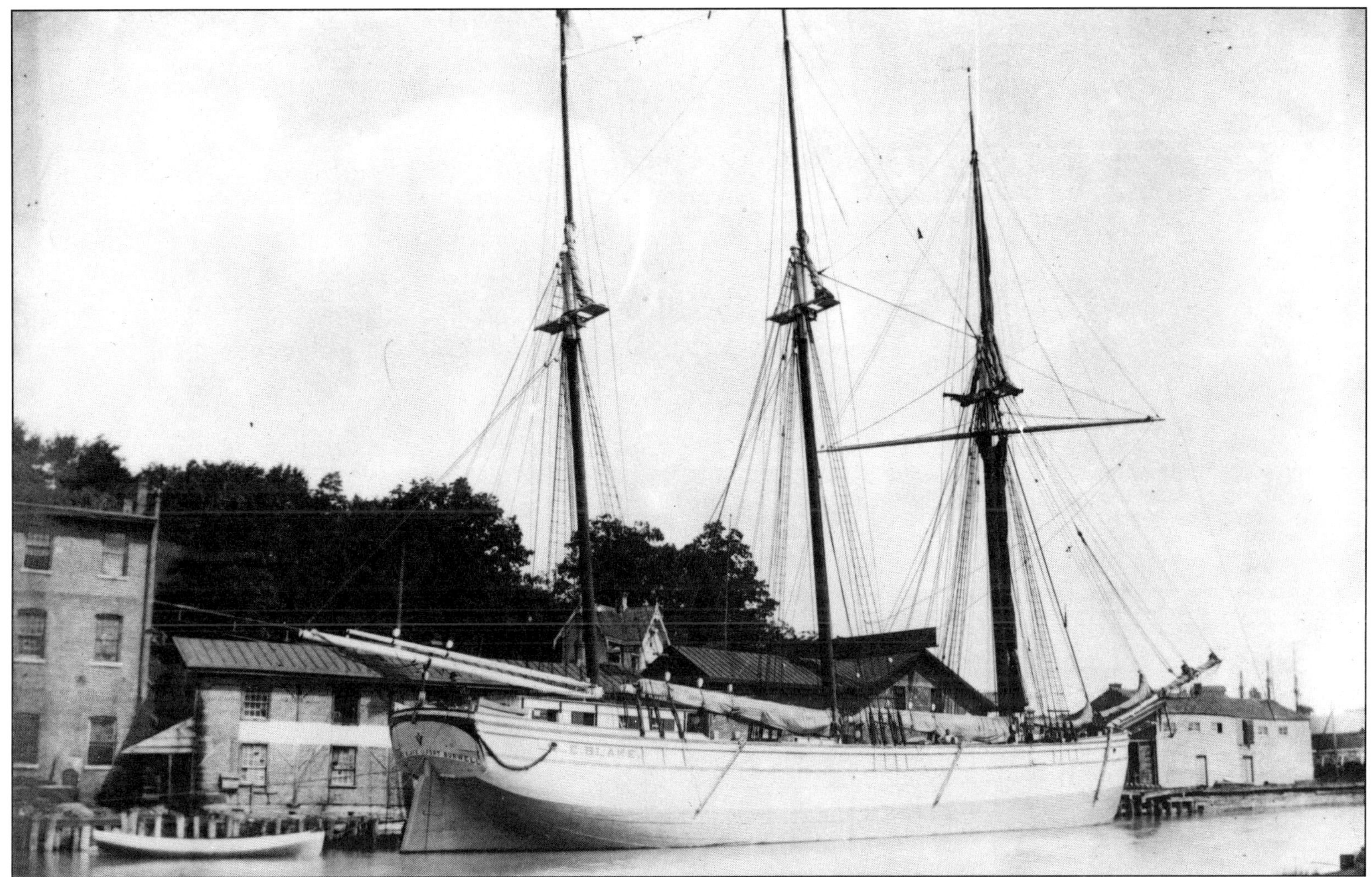

St. Catharines Historical Museum

Edward Blake

Light and ready for loading, the *Edward Blake* reveals the characteristic hull shape of vessels designed for service in the canals and shallow harbours of the Great Lakes. The Lake Erie town of Port Burwell, Ontario, where it sailed from, was such a harbour, and it is seen in this picture lying alongside the canal wharves in St. Catharines, Ontario. The squared-off bow and stern and the almost vertical sides allowed for maximum carrying capacity while fitting into the locks of the Welland and other canals.

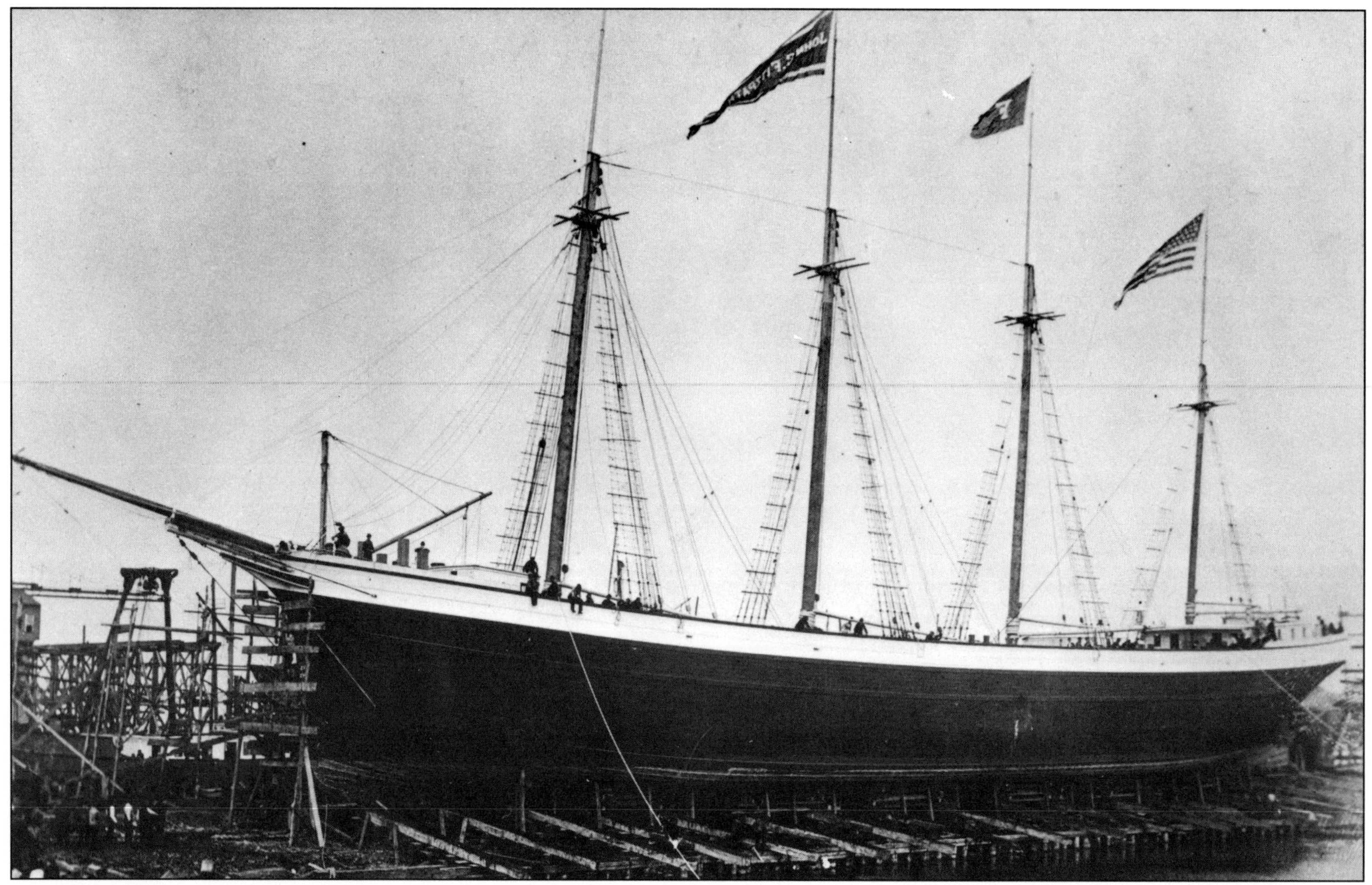

John C. Fitzpatrick

Institute for Great Lakes Research, Bowling Green State University

The four-masted *John C. Fitzpatrick* is shown ready for launching at West Bay City, Michigan, on May 31, 1892. This vessel represented the climax of development of the Great Lakes schooner. Its dimensions represent the maximum-sized vessel capable of transiting the Welland Canal of that era. The lack of external ballast made such vessels difficult to control when light (without cargo).

ORIGIN, DESIGN AND DEVELOPMENT

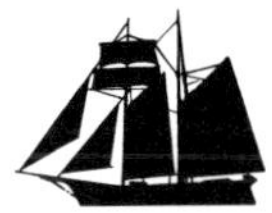

Marine historians have traced the origins of the schooner to the Netherlands in the seventeenth century, as well as Scandinavia and the Mediterranean. Whatever its origin, few will argue that the schooner reached its highest form of development in colonial North America during the eighteenth and nineteenth centuries. The name is said to have come from an old Scottish word meaning skipping over the water. Its first use was recorded at the launching of a ship in New England in the early 1700s.

Great Lakes schooners were similar to their oceangoing counterparts at first. As builders and sailors gained experience specific to the Great Lakes however, the design was modified to reflect the special conditions, to exploit advantages and to accommodate certain constraints.

The hulls of schooners intended to transit the Welland and other canals, for example, were built to have as large a cargo capacity as possible while still fitting into the locks. This made them appear very square or box-like compared to ocean vessels.

Another adaptation was the retractable centreboard. To facilitate sailing close to the wind while not increasing the draught of the vessel beyond the depth of the canals and shallow harbours, a narrow box structure was constructed along the keel. A centreboard mounted on a pivot could be raised into the box while in shallow water and lowered while under sail.

The masts of Great Lakes schooners were proportionately taller than those built for ocean service. Prevailing winds on the lakes were not generally as strong as those at sea and therefore the need to reduce windage caused by extra rigging was not a factor.

The largest schooner ever built was the seven-masted *Thomas W. Lawson* which worked in the Atlantic coasting trades at the turn of the century. Lake schooners ranged in size from quite small, 30 metres (100 feet) or less, to well over 60 metres (200 feet). The five-master *David Dows* and the four-master *Minnedosa* were the largest built on the American and Canadian sides of the Great Lakes respectively (see pp. 52, 54 and 55).

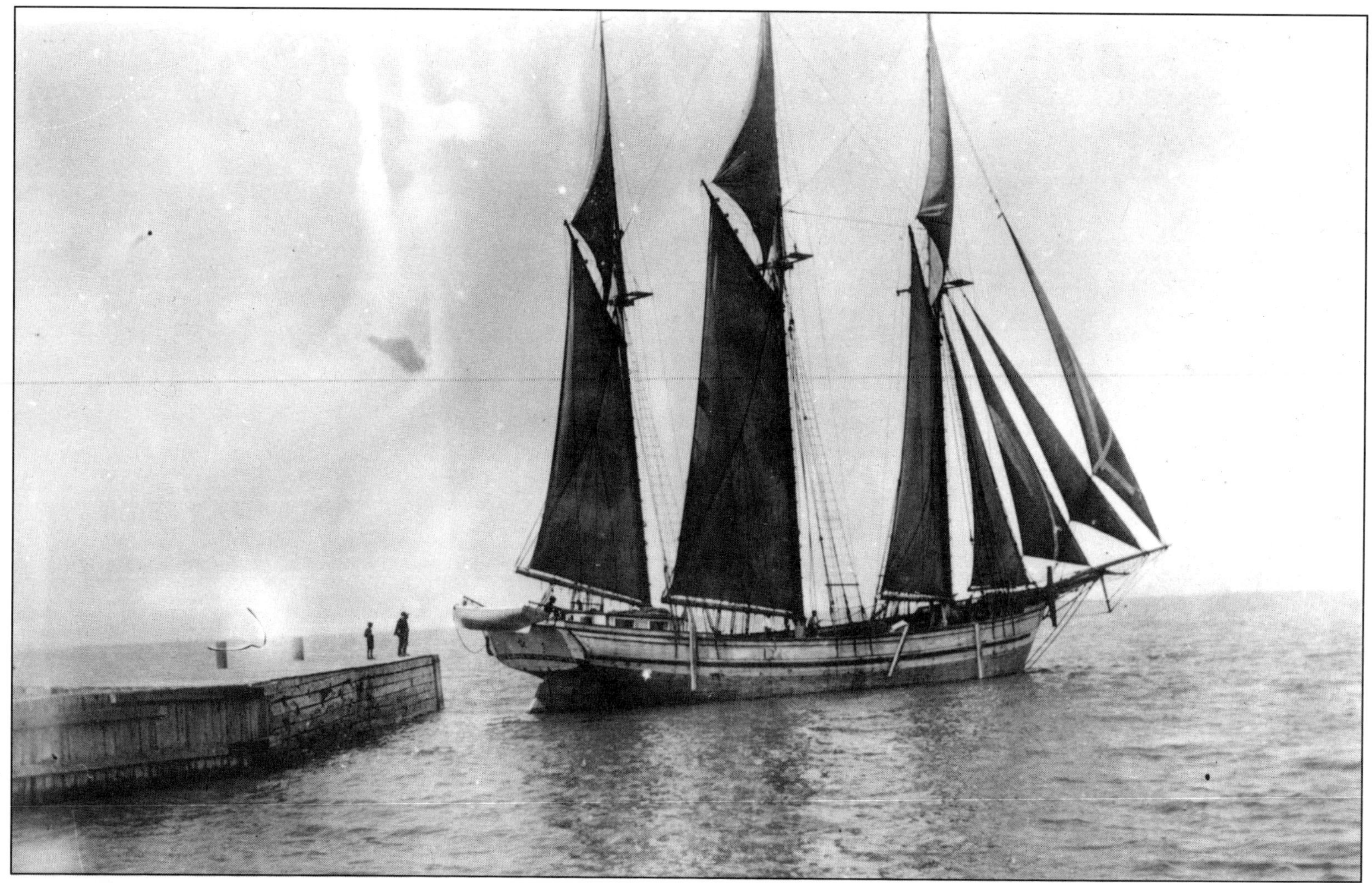

Bert Barnes

National Archives of Canada

The relatively small vessel *Bert Barnes* shows many of the classic characteristics of the Great Lakes schooner. It has the squarish hull, flat stern and tall masts. It probably hailed from the Lake Huron port of Southampton, Ontario, where it would have needed the shallow draught afforded by the retractable centreboard. Only three crewmen can be seen on the deck working the lines as it clears the breakwall of a lake harbour.

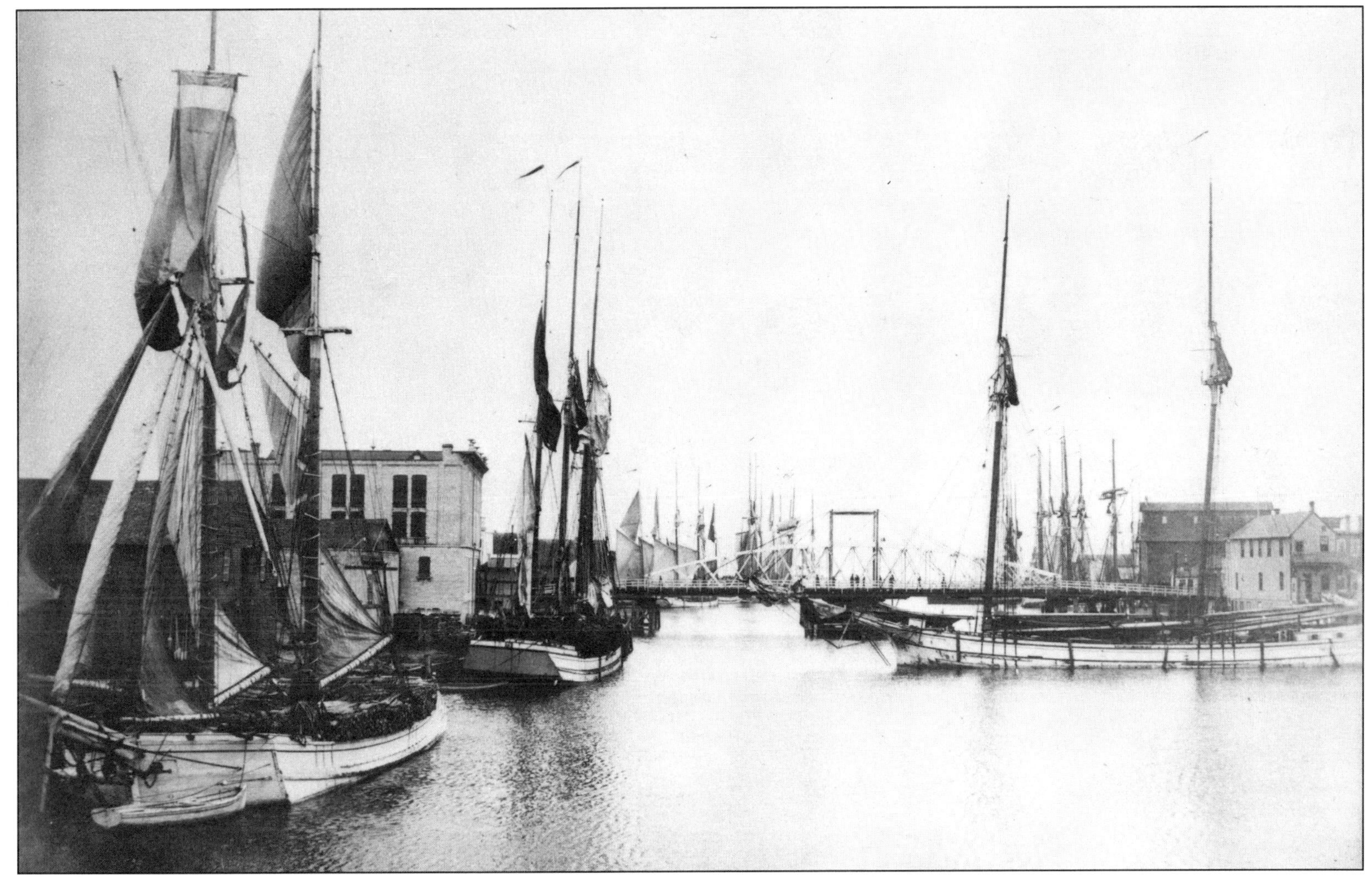

Institute for Great Lakes Research, Bowling Green State University

Behrman, Major Ferry and Mediterranean

Many different variations of the standard Great Lakes schooner design co-existed at any given time during the age of sail. In this 1887 view of the harbour at Manitowoc, Wisconsin, we can see the scow schooner *Agnes Behrman* on the left, the three-masted *Major N. H. Ferry* behind and the two-master *Mediterranean* on the right.

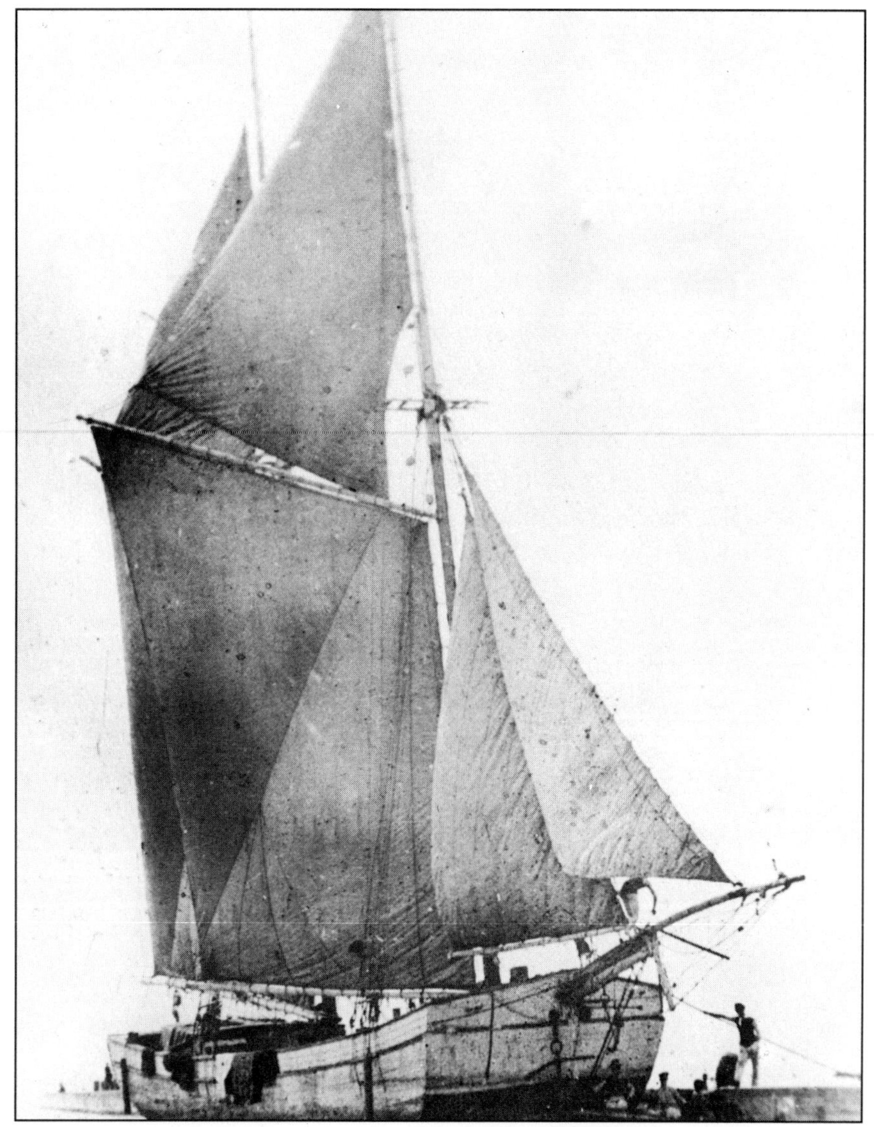

Marine Museum of Upper Canada, Toronto

The Scow Schooner *E. Bailey*

The practicalities of life around the Great Lakes led to design of a unique type of vessel. Even in the late nineteenth century the area was still a frontier and the square hull of the scow was easier for less skilled shipbuilders to construct. By simply putting a schooner sailing rig on top they came up with an inexpensive and useful, if somewhat ungainly looking, craft. This picture shows the scow schooner *E. Bailey* in the eastern gap of Toronto harbour about 1903. Scow schooners were used especially in the Lake Ontario stonehooking fleet where they carried rocks harvested from the lake bottom to be used in construction. Variations on the scow schooner are known to have existed as far away as New Zealand.

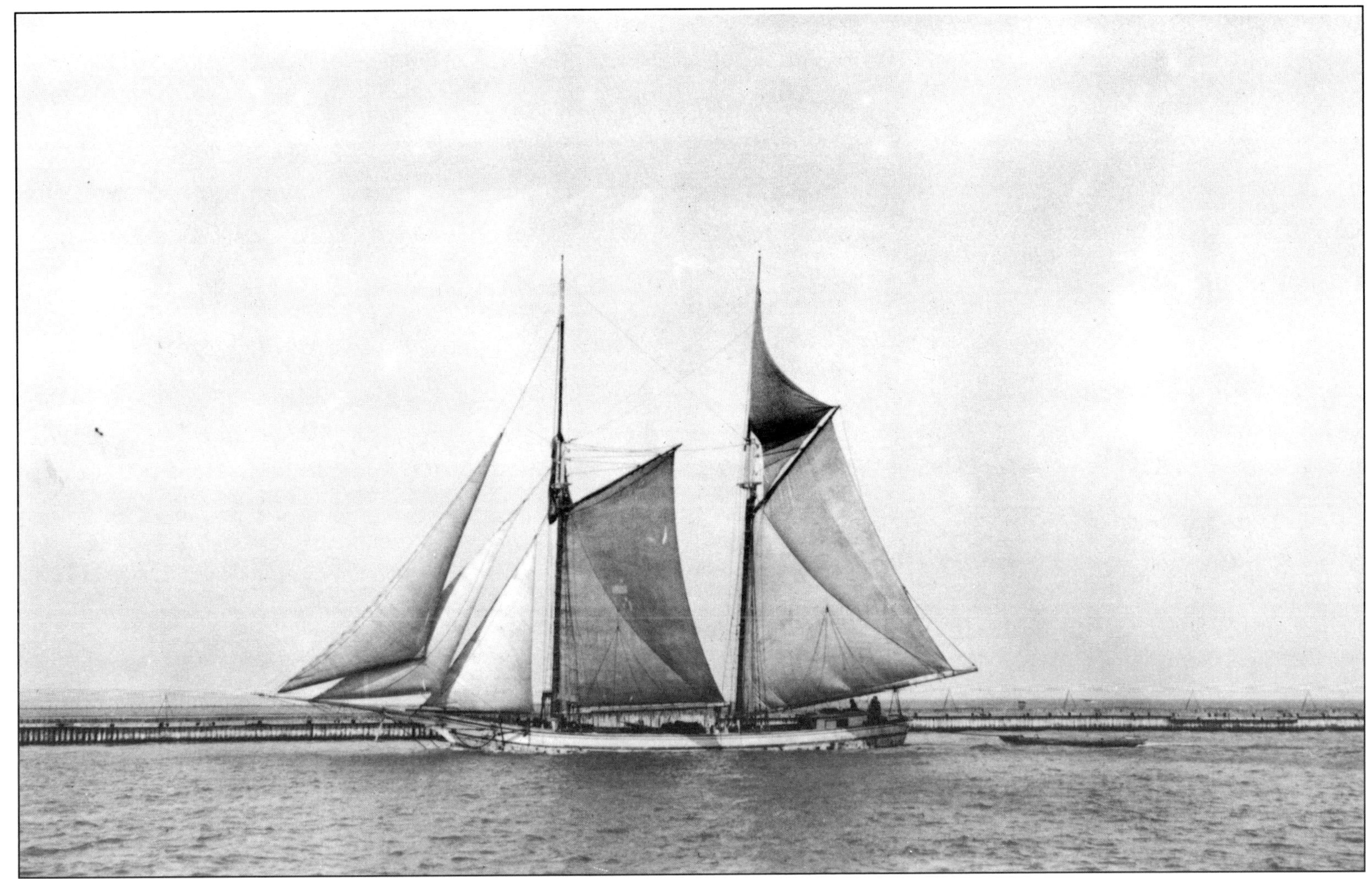

National Archives of Canada

The Stonehooker *Maple Leaf*

Other types of vessels were also engaged in stonehooking such as the conventional two-masted schooner *Maple Leaf*, shown here entering a port, probably in western Lake Ontario. *Maple Leaf* was built in Picton, Ontario, in 1867. The *Maple Leaf* is mentioned in the old sea shanty recalling a collision on Lake St. Clair.

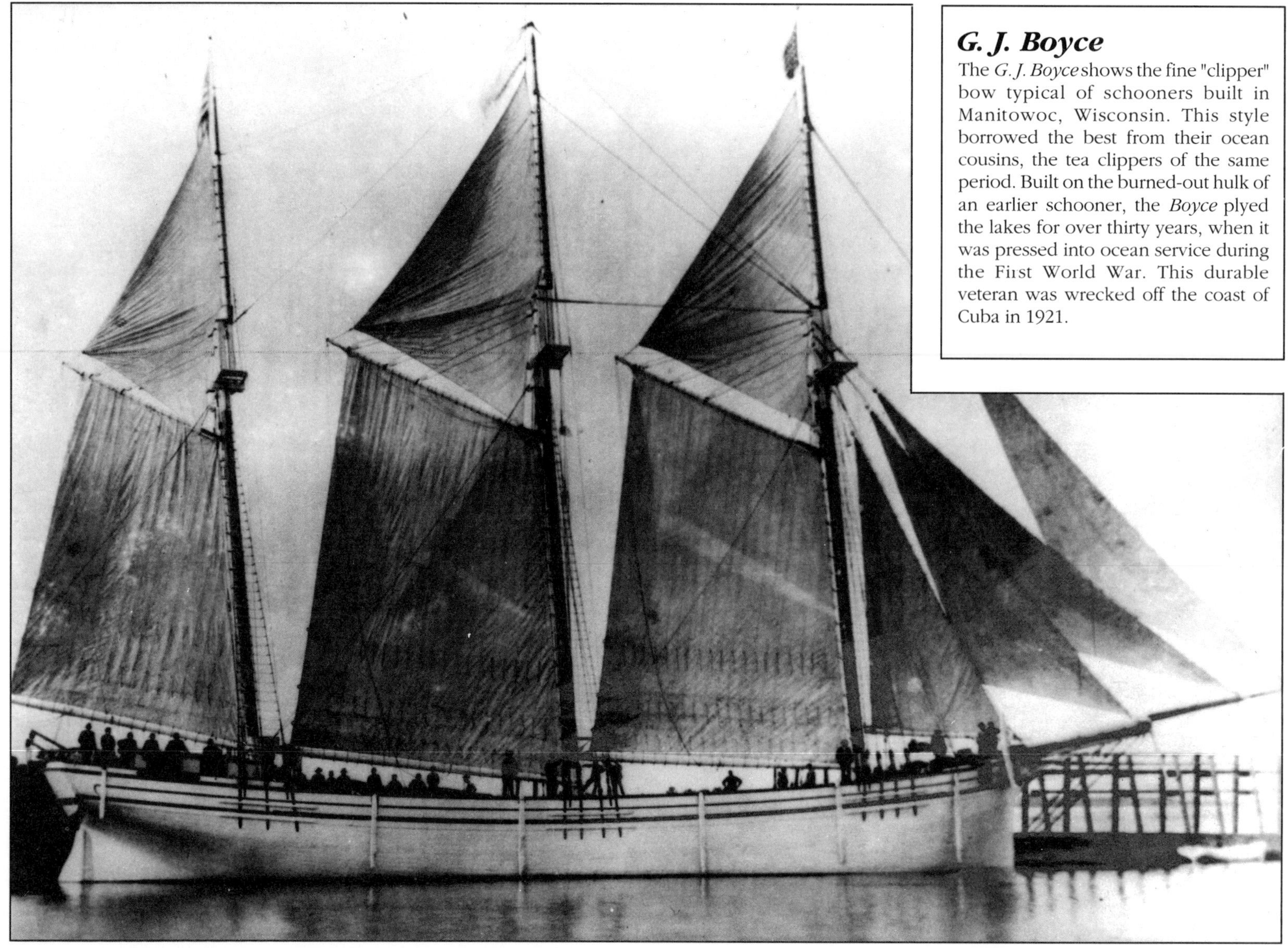

G. J. Boyce

The *G. J. Boyce* shows the fine "clipper" bow typical of schooners built in Manitowoc, Wisconsin. This style borrowed the best from their ocean cousins, the tea clippers of the same period. Built on the burned-out hulk of an earlier schooner, the *Boyce* plyed the lakes for over thirty years, when it was pressed into ocean service during the First World War. This durable veteran was wrecked off the coast of Cuba in 1921.

Manitowoc Maritime Museum, Manitowoc, Wisconsin

THE SAILING RIG

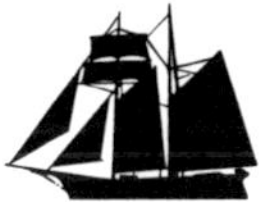

At the height of its development the Great Lakes schooner sailing rig represented one of the most sophisticated and elegant expressions of wind harnessing technology.

Sails were generally of cotton canvas. Many old photos show the patches that indicate long use and extensive maintenance (see *Emma Nielsen* pg. 19). The cordage of both the standing and running rigging was most often hemp rope. These materials were basically the same as those used on ocean vessels, but the less corrosive nature of fresh water allowed the extensive employment of metal, primarily wrought iron, in the rigging. Spars, the collective name for masts, booms, gaffs and yards, were made from the white pine and other well-suited woods plentiful in the Great Lakes forests.

The long bowsprit and jib-boom could accommodate several foresails. These allowed manoeuvrability, and they gave stability when sailing close to the wind. This latter characteristic was extremely important for sailing vessels on the Great Lakes where shoals, islands and narrow channels restricted movement. Sailing close to the wind allowed a vessel to operate even when there was risk of being blown ashore.

Mainsails on each mast were secured to the mast with wooden hoops that allowed them to slide easily up and down. The sails were suspended between a gaff, the spar at the top, and a boom at the bottom. Lines of cords along the lower halves of the mainsails were used to tie or reef the sail when it was partially lowered to reduce its area in strong winds. The mainsails were controlled by lines called braces that attached to their gaffs and sheets that secured their booms. Triangular topsails were set between the main gaffs and the topmasts when the vessel were likely to be sailing longer distances. While entering harbour or manoeuvring, they were apt to be left furled (see *R. P. Mason*, pg. 20).

One totally unique feature of the Great Lakes schooner rigging was the triangular fore-topsail known as a raffee (see *Lizzie Metzner*, pg. 18).

The attractiveness of the Great Lakes schooner rig in terms of its operational cost was that all the sails could be managed by only three or four sailors. They could be raised, lowered and reset by manipulating lines and halyards that could all be reached from the deck. Only on rare occasions did a sailor have to go aloft. In later years a steam-powered donkey engine was added to the foredeck of many schooners. Through pulleys and tackles most lines could be hauled using this device, reducing the work of the sailors even more (see *Emerald*, pg. 58).

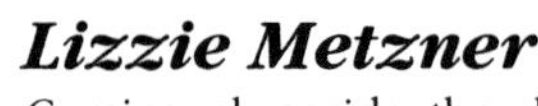

Lizzie Metzner

Coming alongside the docks in the Manitowoc River, Wisconsin, in a light breeze, the small schooner *Lizzie Metzner* shows the full complement of sails typical of its class. Built in 1888 at Manitowoc by Rand and Burger, the *Metzner* worked the timber trade in Lake Michigan until 1904 when it was sold to Joseph Fagan of Belleville, Ontario, and moved to Lake Ontario. The vessel met its end in 1916 when its sails finally failed and it was driven ashore near Oswego, New York.

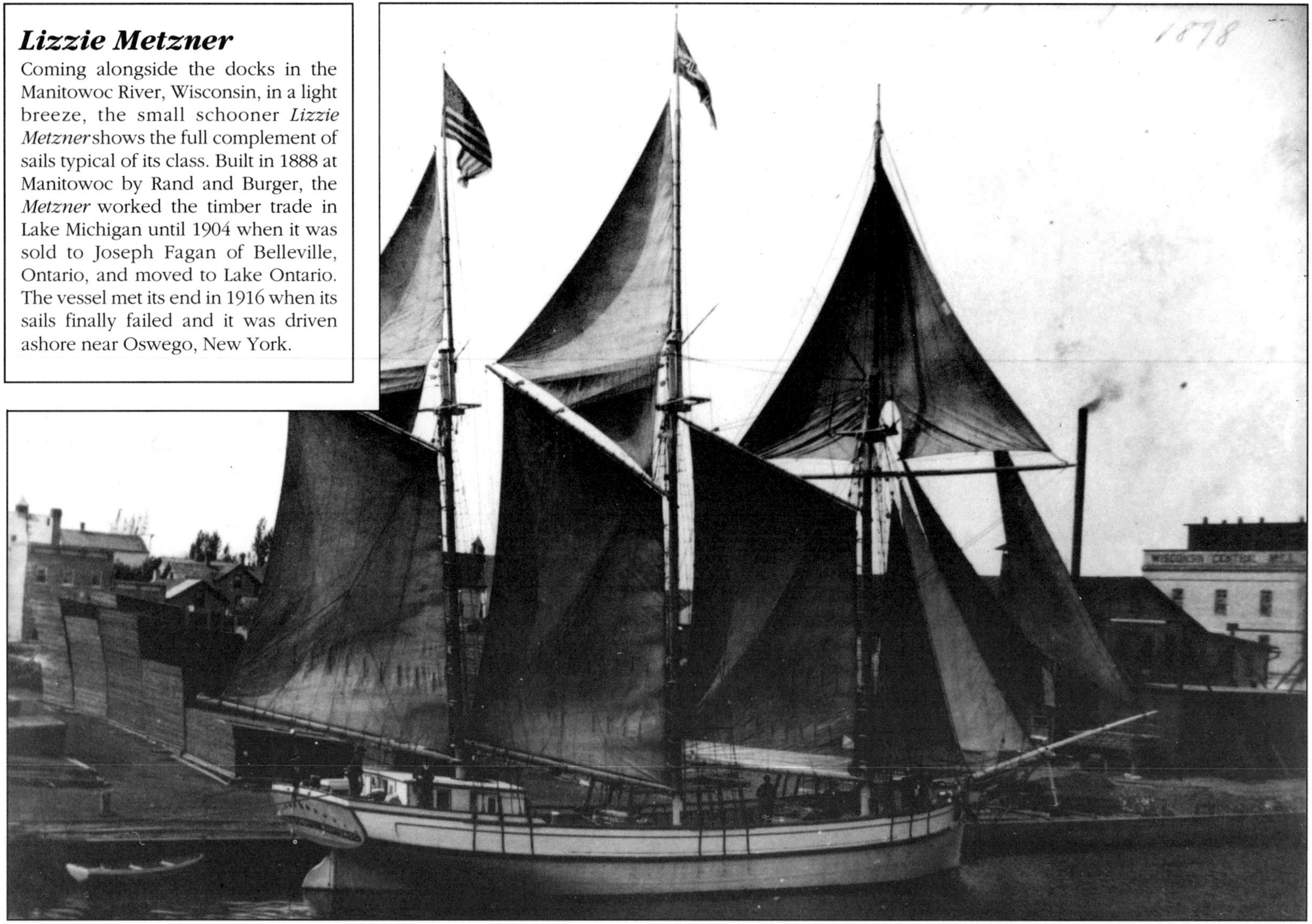

Manitowoc Maritime Museum, Manitowoc, Wisconsin

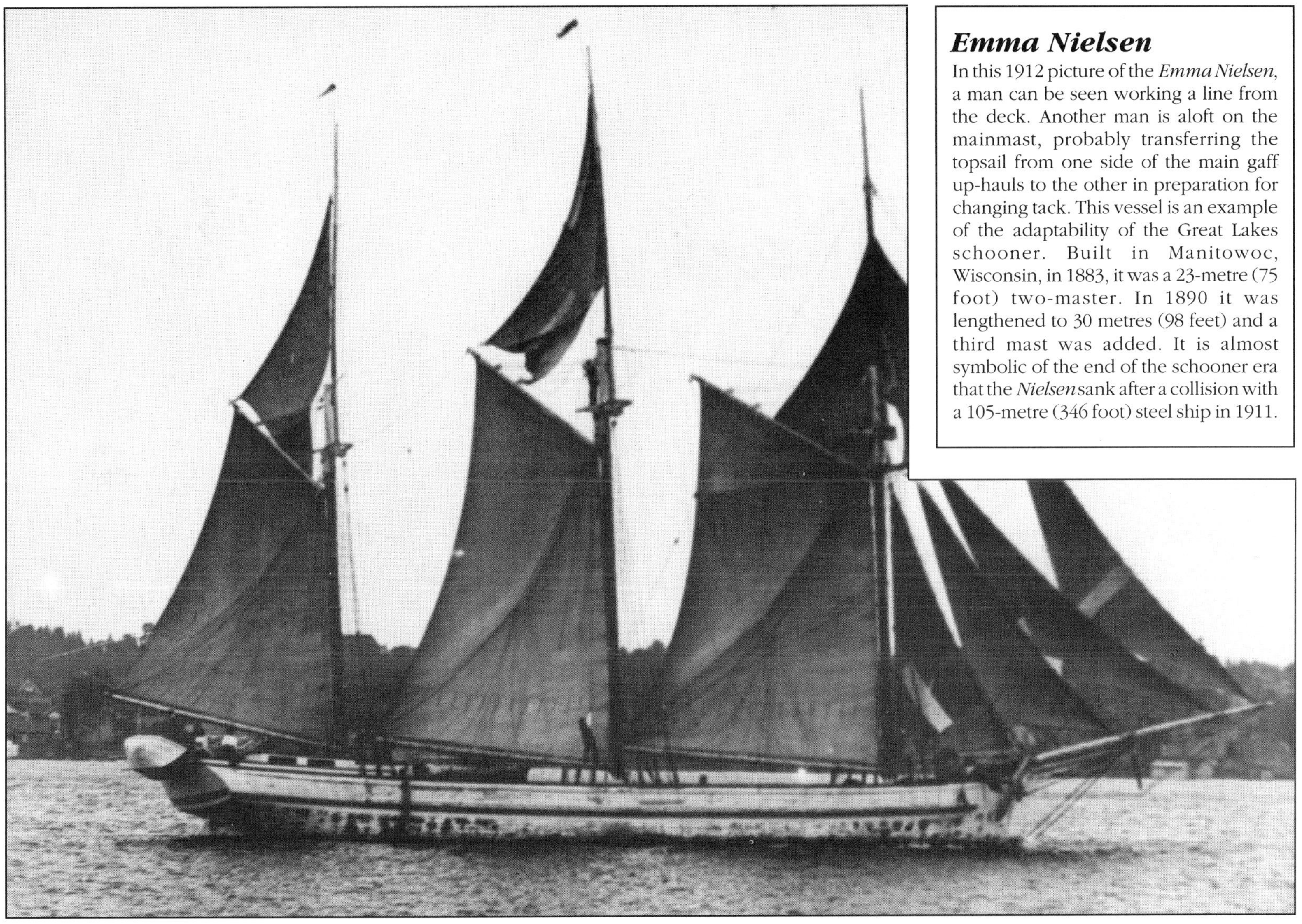

Emma Nielsen

In this 1912 picture of the *Emma Nielsen*, a man can be seen working a line from the deck. Another man is aloft on the mainmast, probably transferring the topsail from one side of the main gaff up-hauls to the other in preparation for changing tack. This vessel is an example of the adaptability of the Great Lakes schooner. Built in Manitowoc, Wisconsin, in 1883, it was a 23-metre (75 foot) two-master. In 1890 it was lengthened to 30 metres (98 feet) and a third mast was added. It is almost symbolic of the end of the schooner era that the *Nielsen* sank after a collision with a 105-metre (346 foot) steel ship in 1911.

Manitowoc Maritime Museum, Manitowoc, Wisconsin

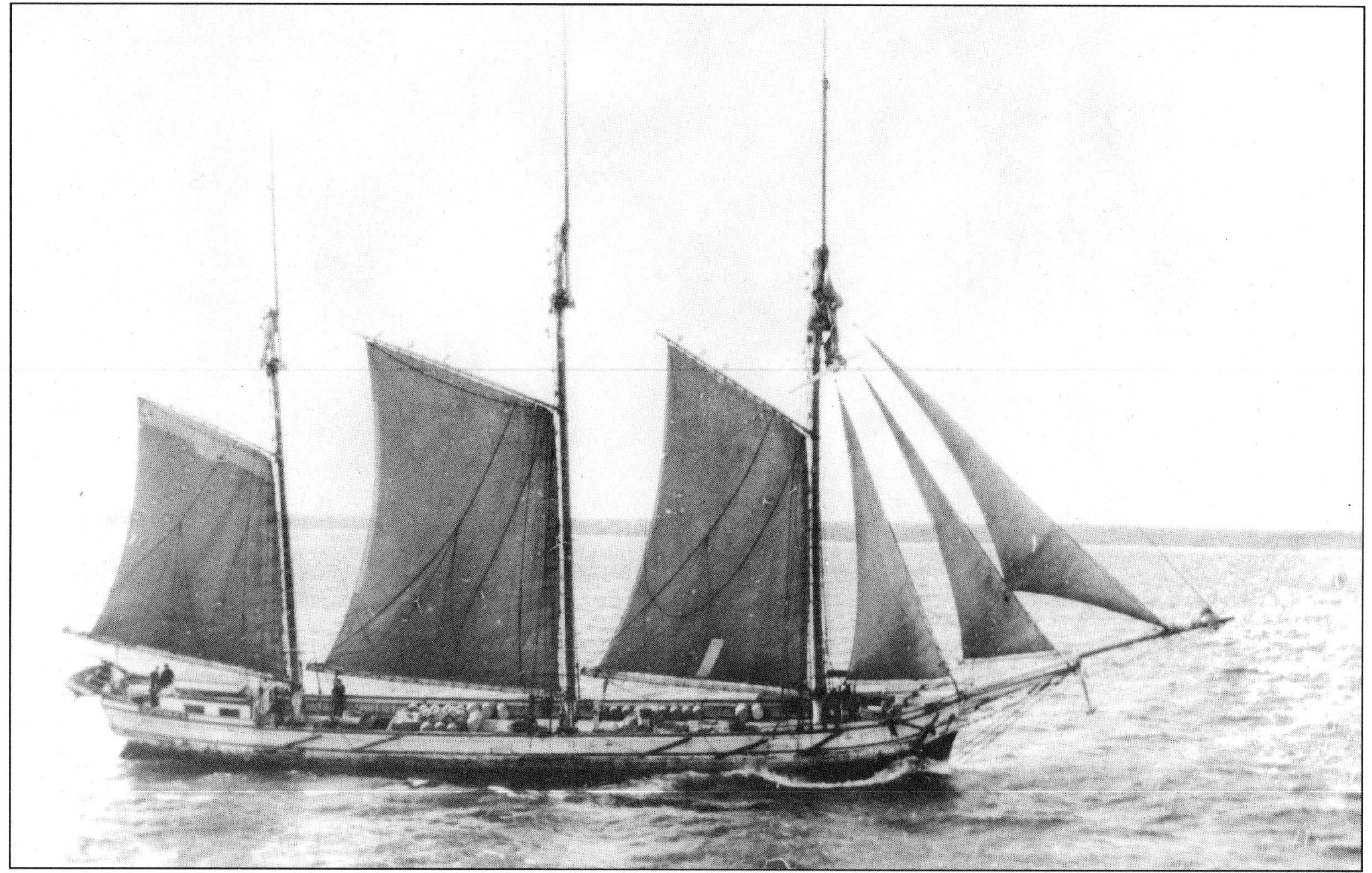

R. P. Mason

Dossin Museum of the Great Lakes, Detroit

The *Mason* is seen underway with a bone in her teeth without its topsails or raffee and flying jib. Schooners did not always use all their sails, but used what they needed under the prevailing conditions. On the deck of the *Mason* can be seen a number of barrels. Especially in the earliest days of Great Lakes waterborne trade many items from flour to nails were carried in barrels. A barrel was a convenient packaging device, since it could be handled by a single man rolling it along a ramp, wharf or deck.

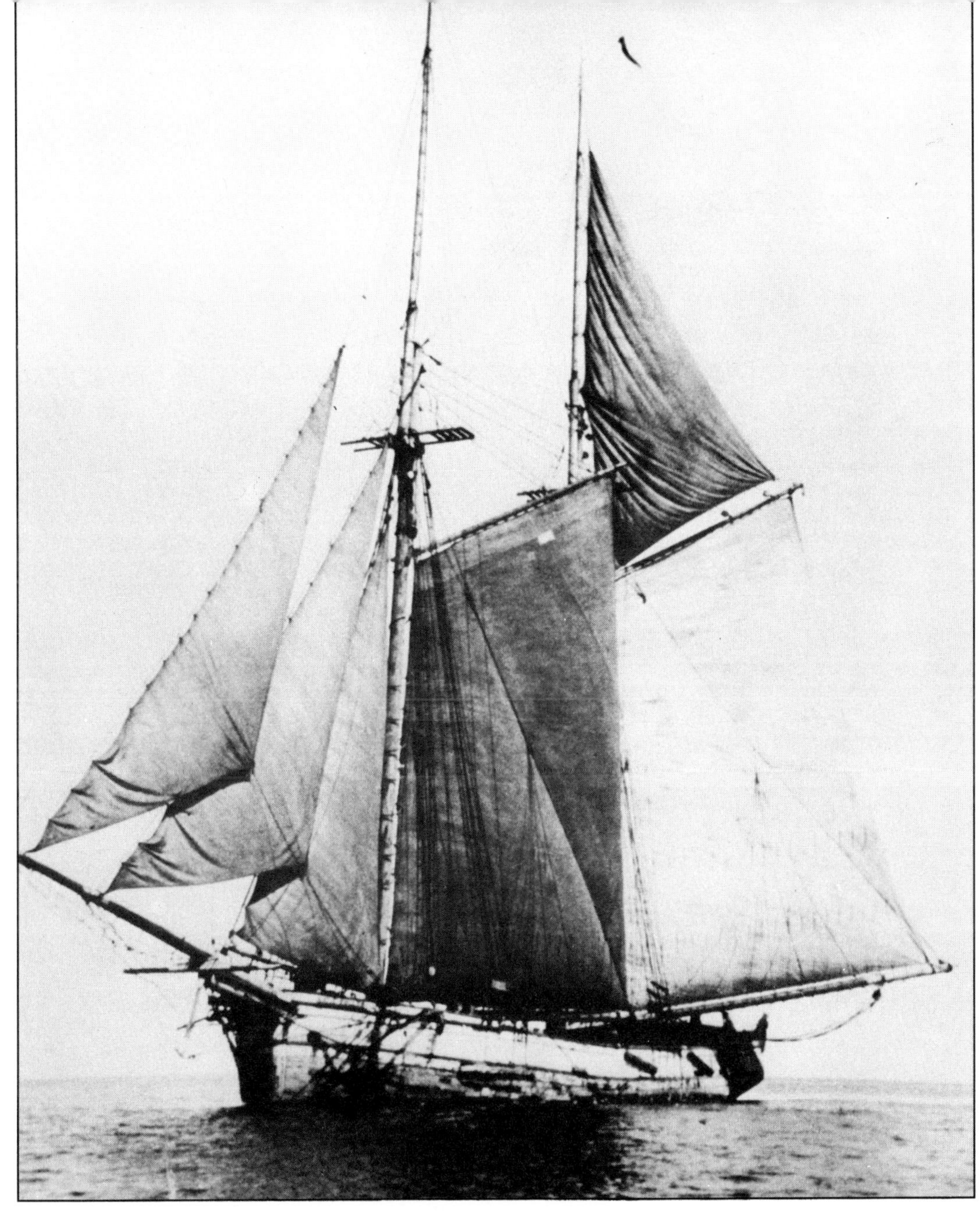

Katie Eccles

The heavily laden *Katie Eccles* is seen probably approaching Port Dalhousie at the northern end of the Welland Canal. The sheet or line holding the mainsail boom is hanging loose indicating little wind. The characteristic tall masts of Great Lakes schooners are most evident. The small flag on the mainmast is a telltale that indicated wind direction to the helmsman. The *Eccles* was built in Deseronto, Ontario, in 1877.

St. Catharines Historical Museum

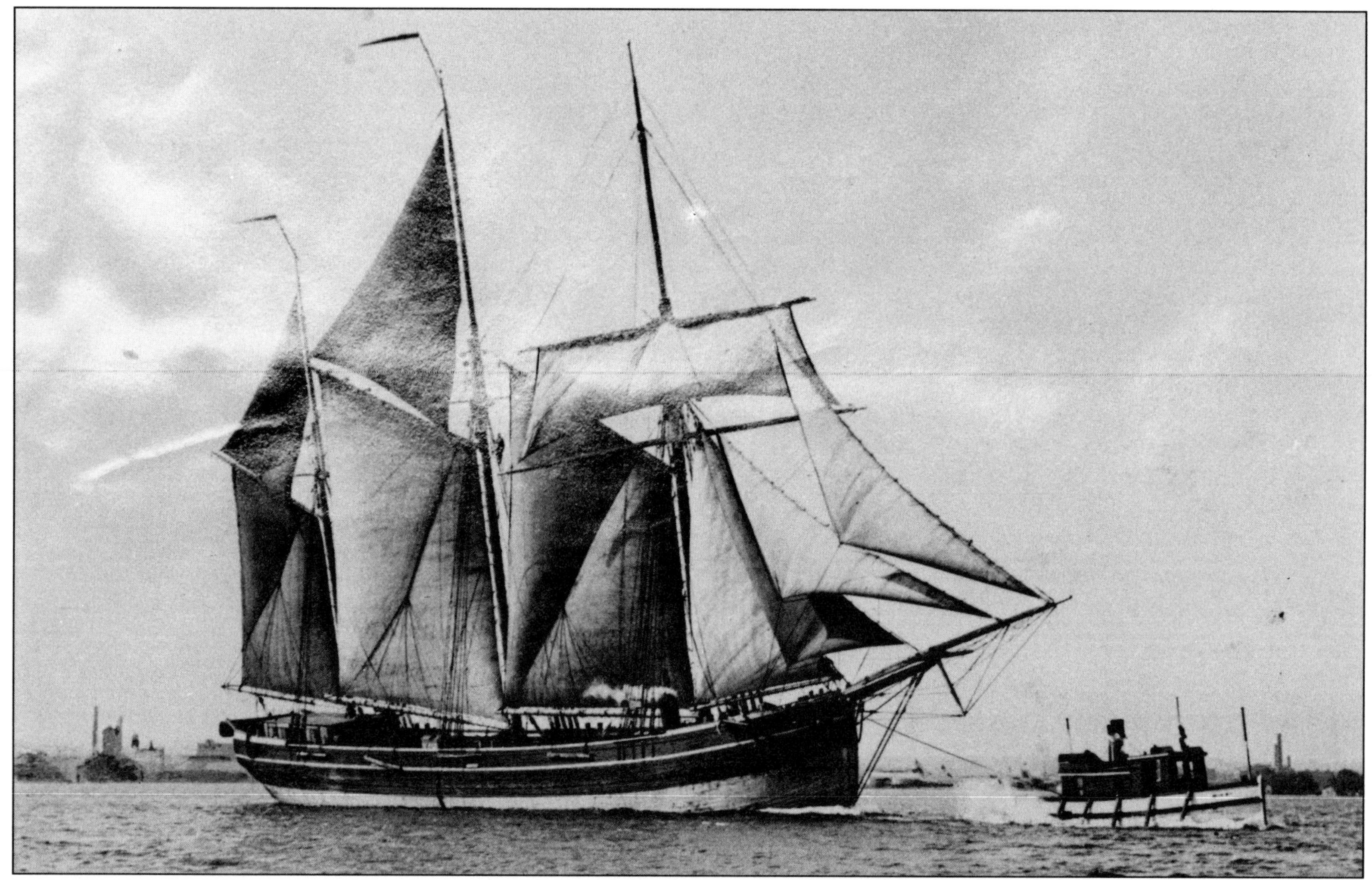

Marine Museum of the Great Lakes at Kingston

Stuart H. Dunn

With the square topsail reefed and the fore gaff-topsail struck, the handsome schooner *Stuart H. Dunn* is towed out of port preparing for another voyage. The *Dunn* was built in 1877 at South Bay in Prince Edward County, Ontario. It was reduced to a schooner-barge in 1910, sailed until 1914 and was stripped and sunk in Lake Ontario in 1925.

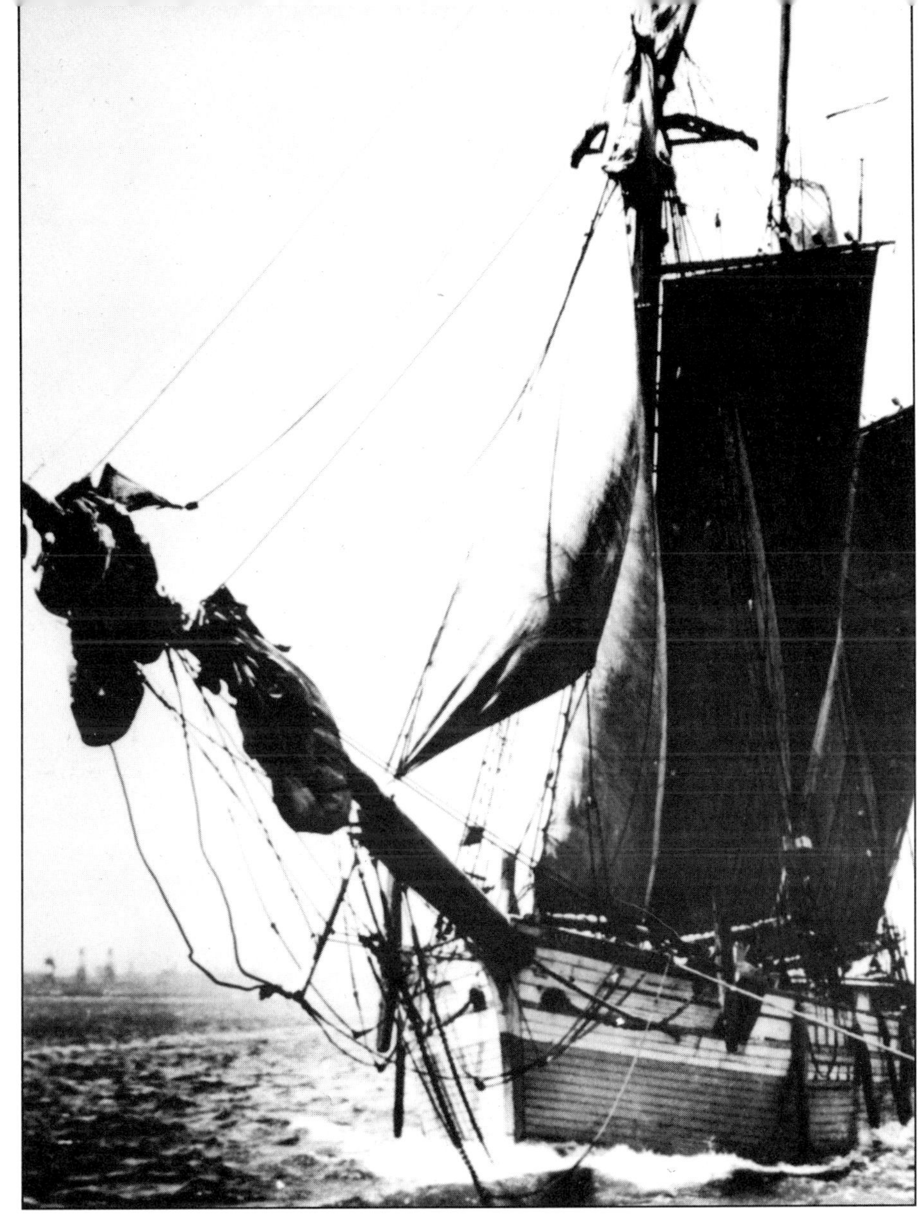

Lucia Simpson

The *Lucia Simpson*, like so many Great Lakes ships, was named after a woman—in this case, the wife of the businessman who financed the construction at Manitowoc, Wisconsin, in 1875. Built of oak at a length of 39 metres (130 feet), it was a member of the medium-large class of schooner. In this picture it is under tow, and since the flying jib and fore-topmast staysail are not set, sheets or lines, being under no tension, are hanging loose (see also pg. 30).

Great Lakes Historical Society, Vermilion, Ohio

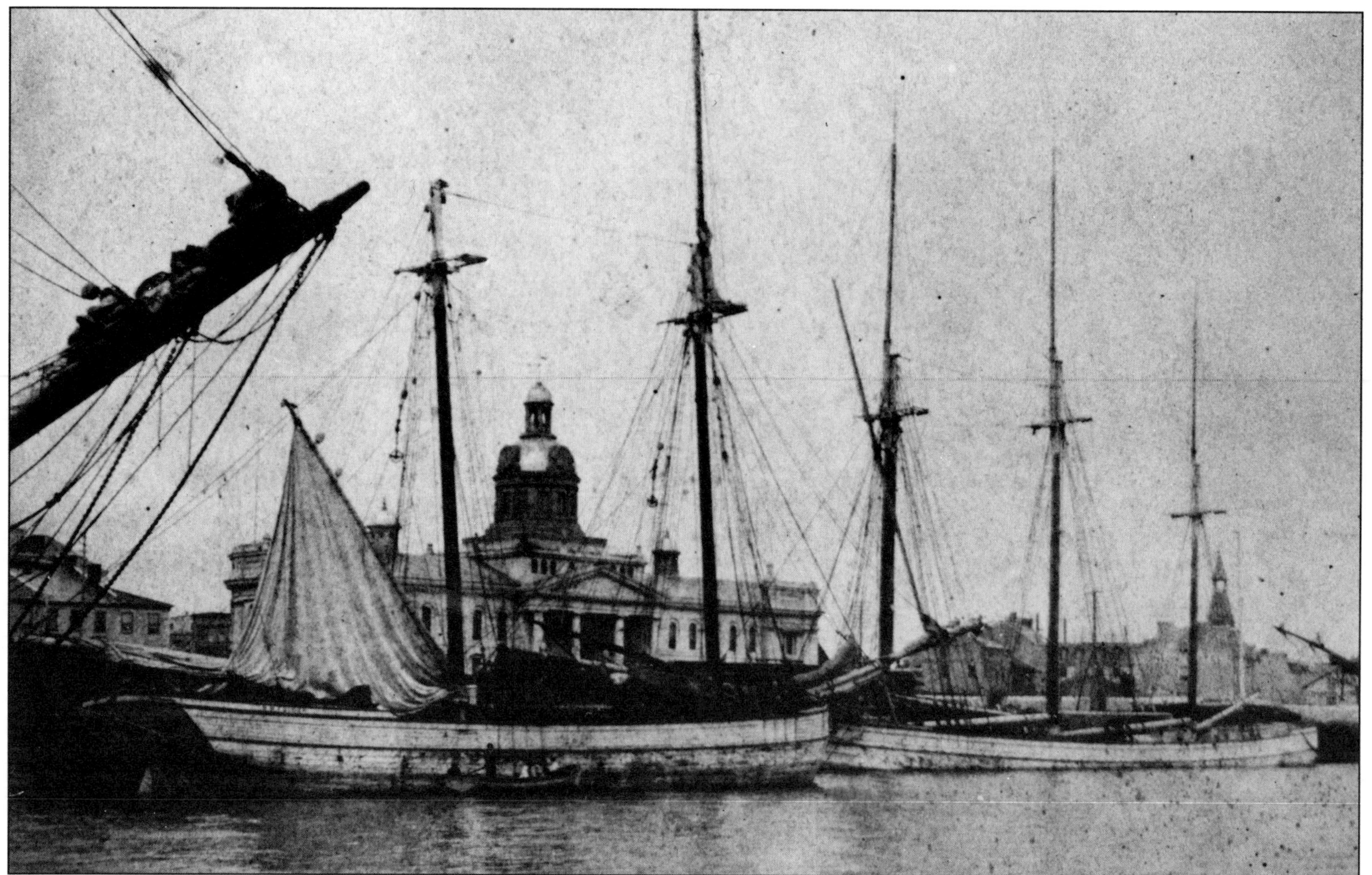

B. W. Folger

National Archives of Canada

The two-masted schooner *B. W. Folger* is pictured in Kingston harbour at dock in front of the old city hall that still graces that city's waterfront. Nowadays yachts and pleasure boats occupy this area. The *Folger*'s mainsail is lowered without the benefit of lazy jacks (see pg. 29), and we can see how the sail drapes onto the deck. Above the mainsail gaff, and also above the foremain, large metal pulleys have replaced the traditional wooden blocks. These types of fittings were more common on freshwater vessels than on their saltwater cousins.

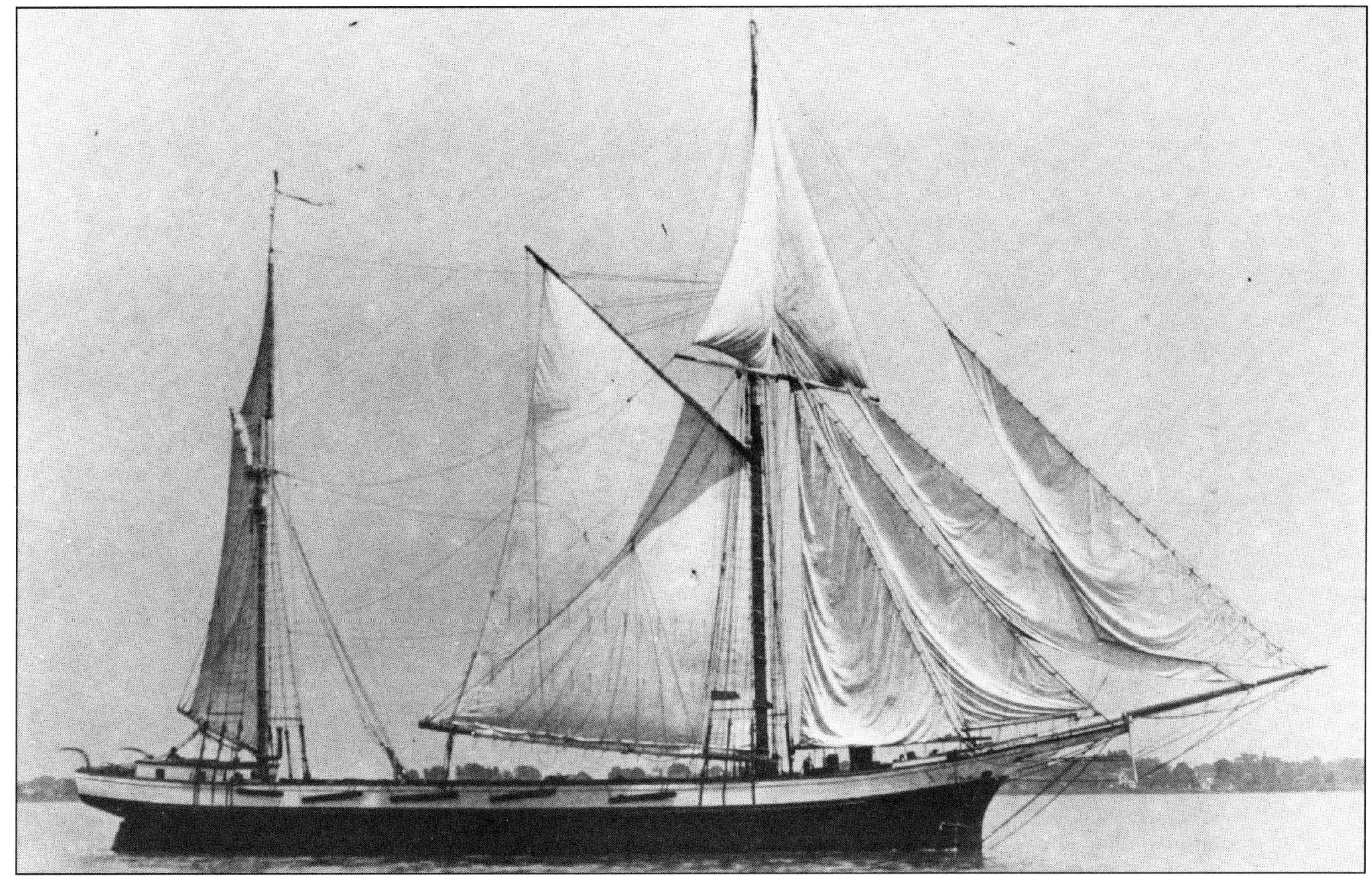

Institute for Great Lakes Research, Bowling Green State University

Jennie Weaver

A variation on the standard two-masted Great Lakes schooner sail pattern was the so-called Grand Haven rig. In this configuration the foremast was taller than the after or mizzenmast. The *Jennie Weaver* is shown sailing wing on wing, that is with its main sail booms extended out opposite sides. This could only be done when the wind was following (coming directly from astern).

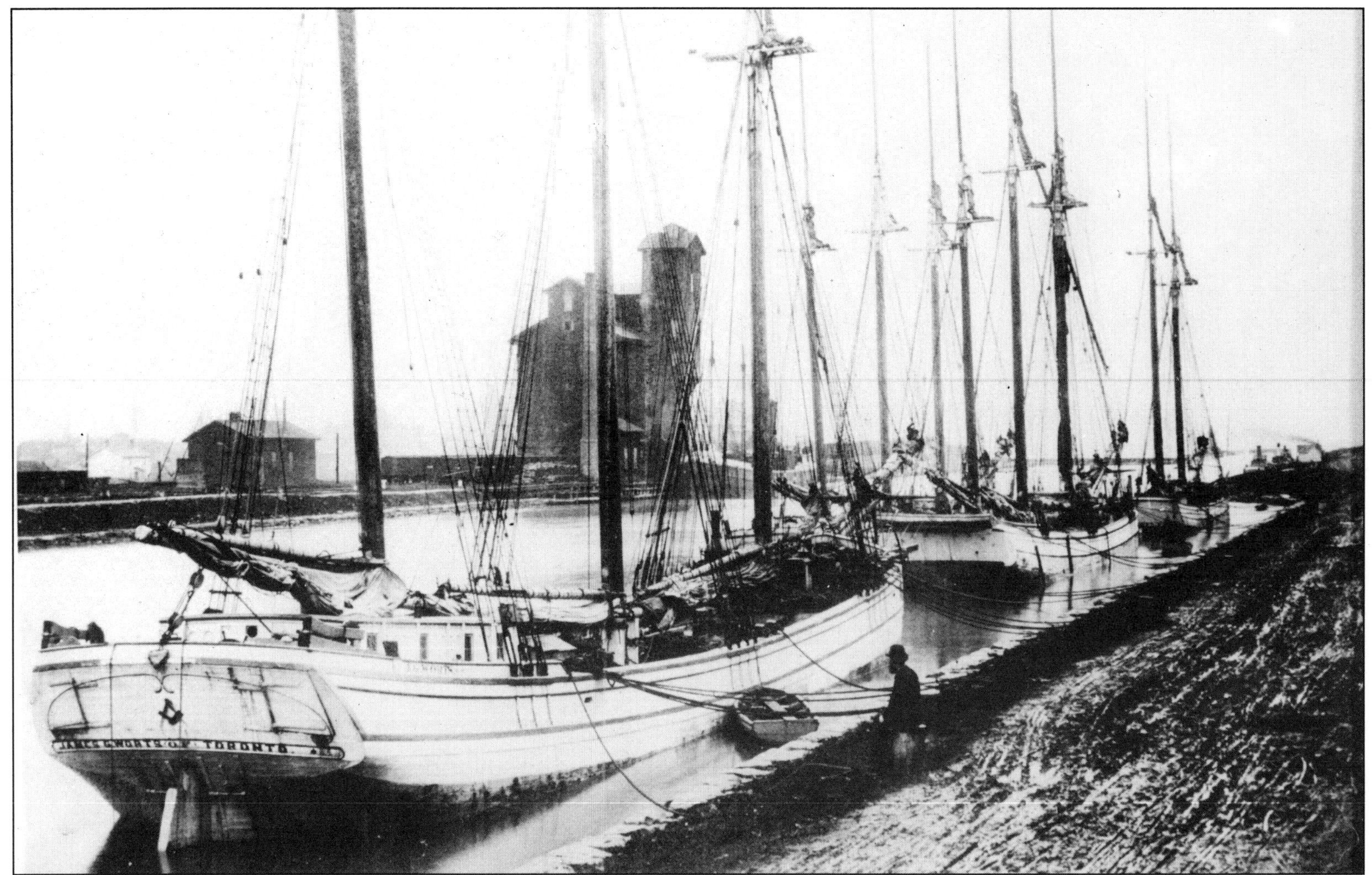

St. Catharines Historical Museum

James G. Worts

This vessel was built in Deseronto, Ontario, on the north shore of Lake Ontario. Owned by the Worts family of the Gooderham and Worts Distillery in Toronto, it was probably engaged in the grain trade relating to the production of whiskey and other spirits. However, evident on the transom are the hatches or loading ports through which long lengths of timber, which did not fit through the deck hatches into the hold, were loaded (see also pg. 51). Also on the transom can be seen the masonic symbol indicating the affiliation of the owners and perhaps the builder (see pp. 28 and 33).

TRADES AND CARGOES

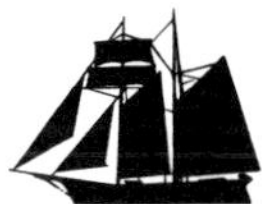

In the earliest days of Great Lakes navigation, before the building of the Erie, Soo, Welland and St. Lawrence canals, schooners were mainly employed collecting furs from far flung trading posts and supplying military forts. As settlers flooded west after the War of 1812, however, they cleared land for agriculture and began to harvest the rich forests. The nature of cargoes began to change fundamentally.

Schooners continued to carry goods such as manufactured articles and food stuffs. Barrels were extremely important in the early stages. Flour, apples, fish, nails and other products were packed in barrels that could be manipulated and rolled along piers and decks by one man. But after the middle of the nineteenth century this trade shifted primarily to trains. It was bulk cargoes that became the staple of the sailing fleet.

The holds of ships were designed more and more to receive cargoes such as grain and coal directly. Timber was often loaded through special hatches or ports, at both bow and stern. Grain was collected in large elevators from the hinterland surrounding ports. It went directly out through the Welland and St. Lawrence canals to foreign markets or was trans-shipped at Buffalo to the smaller Erie Canal barges. Timber also went directly abroad or to ports on the southern boundaries of the lakes for rail shipment east and west. Coal came by canal barge or train from the mines of Appalachia to Lake Erie ports for distribution to the growing industrial centres of the mid-western United States and Ontario.

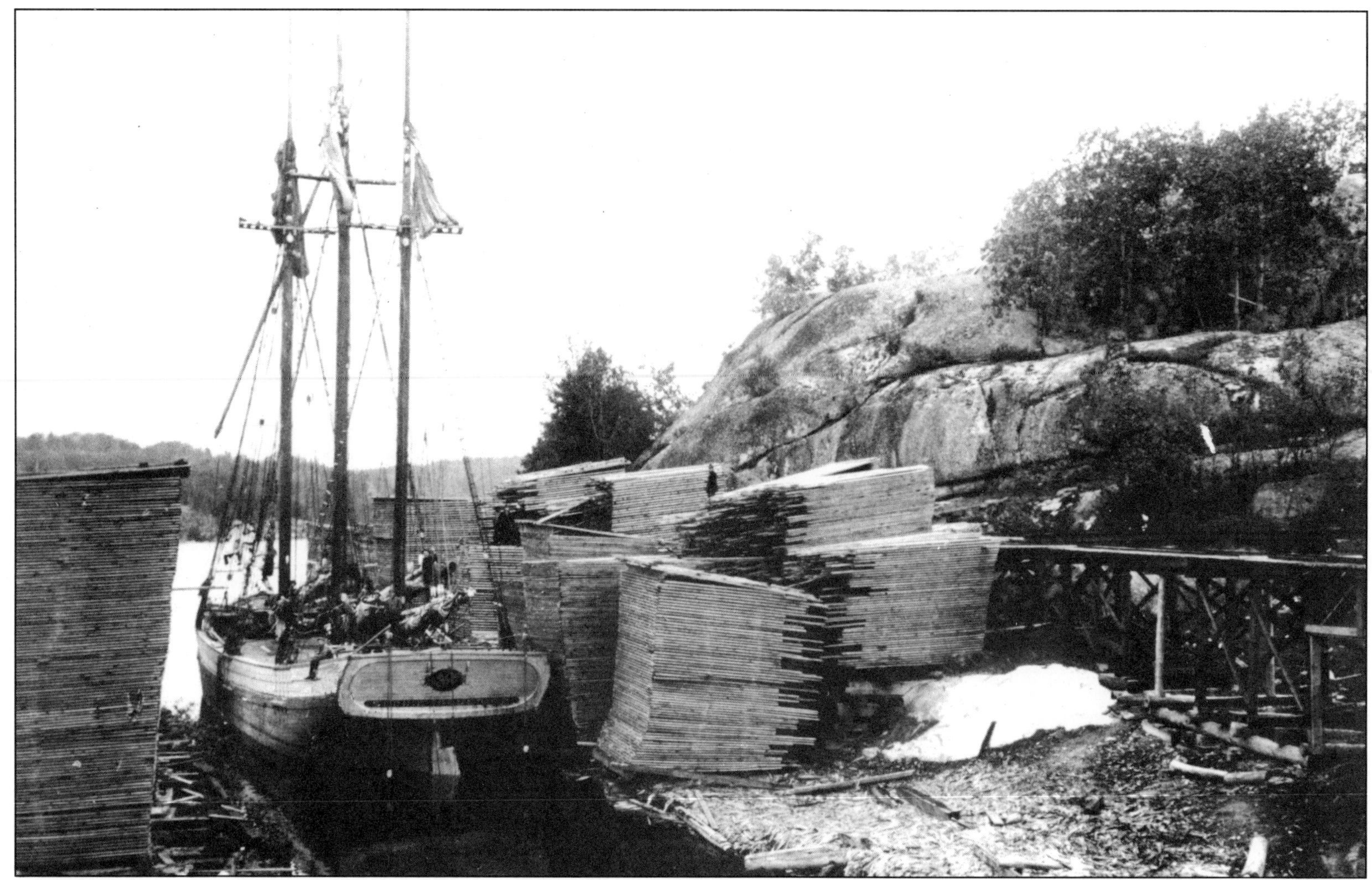

Sephie

Marine Museum of the Great Lakes at Kingston

The *Sephie*, built in Goderich, Ontario, reflects the beautiful lines of the schooner. Its ornate decorative transom was not untypical of workaday ships on the lakes (see pp. 26 and 33). Shown ready to load lumber at Collins Inlet on Manitoulin Island, the *Sephie* traded on the lakes until sold in 1916.

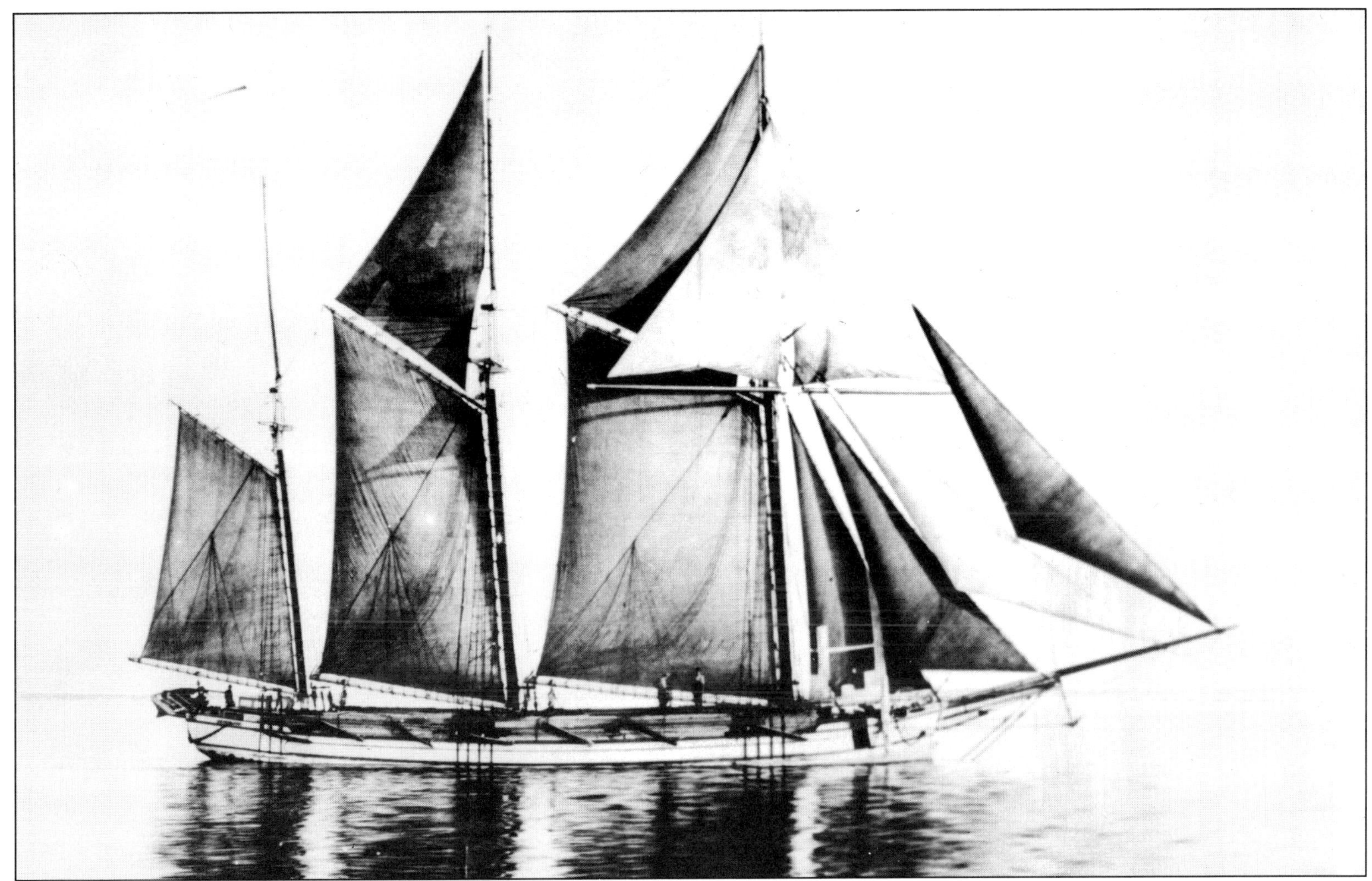

Great Lakes Historical Society, Vermilion, Ohio

J. B. Newlands

Heavily laden with sawn lumber on the deck, and presumably in the hold, the *J. B. Newlands* is seen moving smartly in a fresh breeze on Lake Ontario. It was built in Manitowoc, Wisconsin, in 1870 and served until the 1920s when it was scuttled off Kingston, Ontario. The lines that can be seen coming up from the booms to a point in the middle of the mainsails are lazy jacks, whose purpose it was to gather the sails when they were lowered.

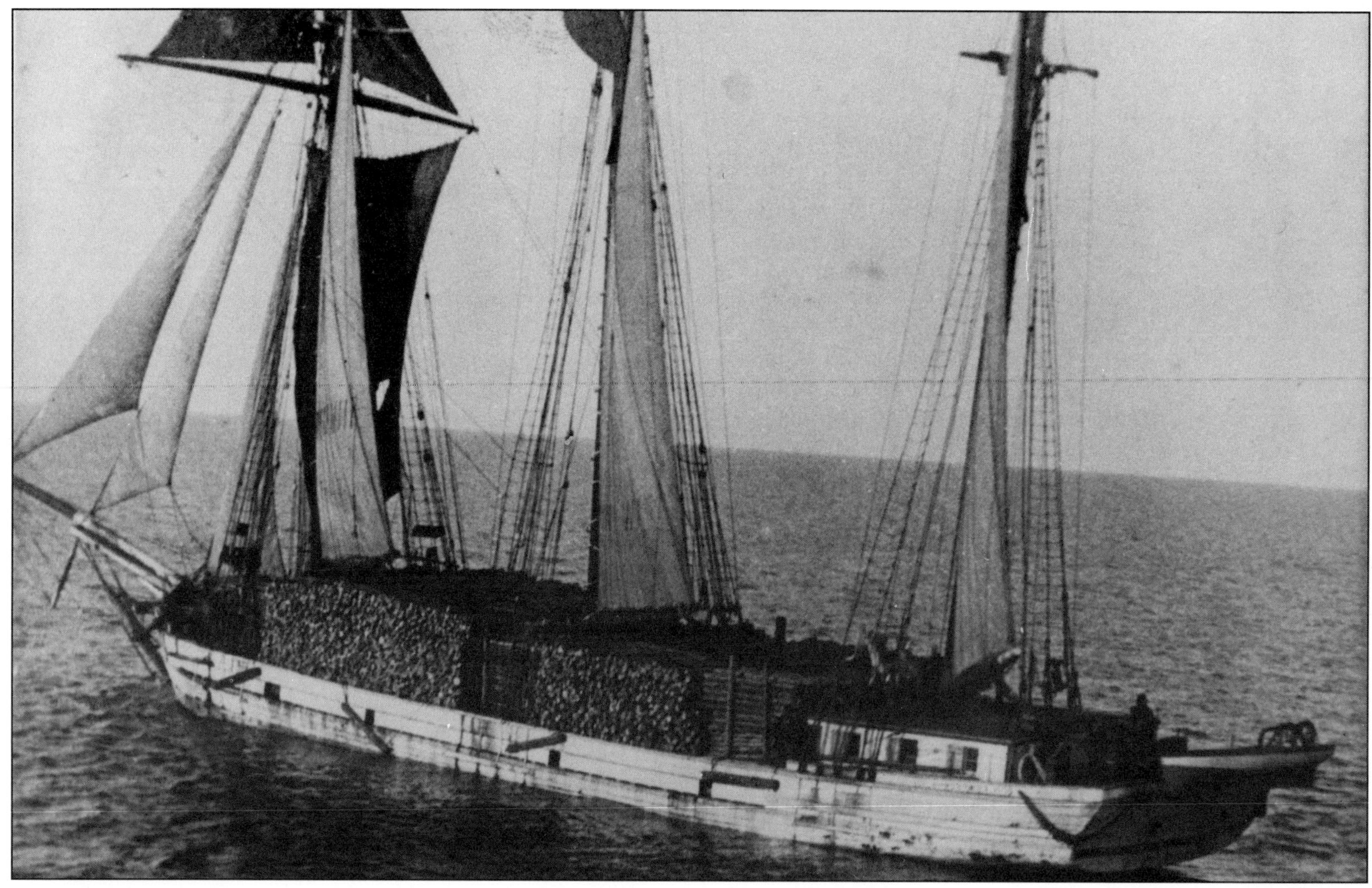

Manitowoc Maritime Museum, Manitowoc, Wisconsin

Lucia Simpson

The considerable capacity for deck cargo, especially lumber, can be seen clearly in this picture. The *Lucia Simpson* (see also pg. 23) could accommodate 350,000 board feet of timber, but was also reported to be a fast sailor capable of 16 miles an hour in its prime. After twenty-seven years in the Lake Michigan timber trade, it was fitted out for salvage work in 1929. Having had little success in that role and after missing a chance to become a floating clubhouse for some Chicago yachtsmen, it was destroyed by fire at Sturgeon Bay, Wisconsin, in 1935.

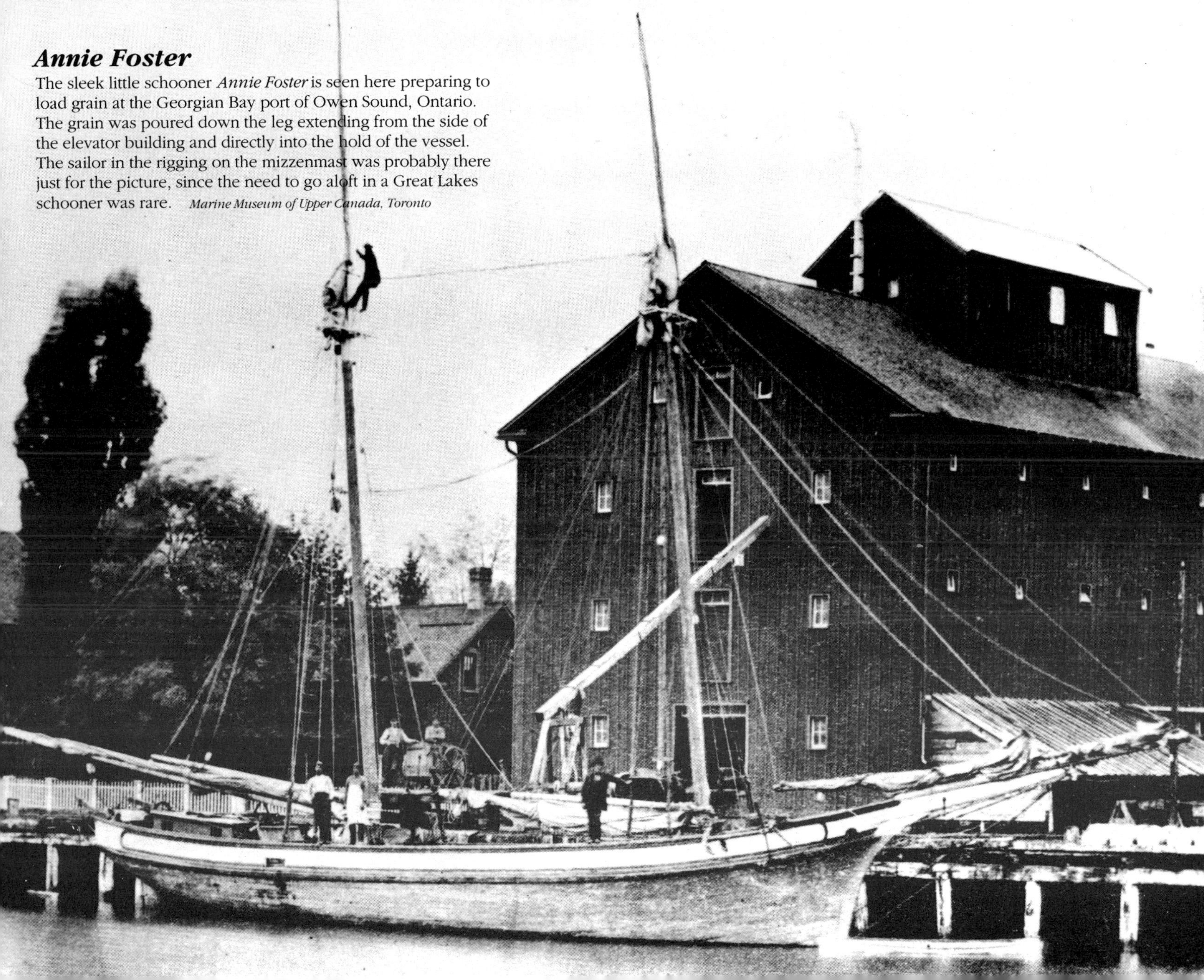

Annie Foster

The sleek little schooner *Annie Foster* is seen here preparing to load grain at the Georgian Bay port of Owen Sound, Ontario. The grain was poured down the leg extending from the side of the elevator building and directly into the hold of the vessel. The sailor in the rigging on the mizzenmast was probably there just for the picture, since the need to go aloft in a Great Lakes schooner was rare. *Marine Museum of Upper Canada, Toronto*

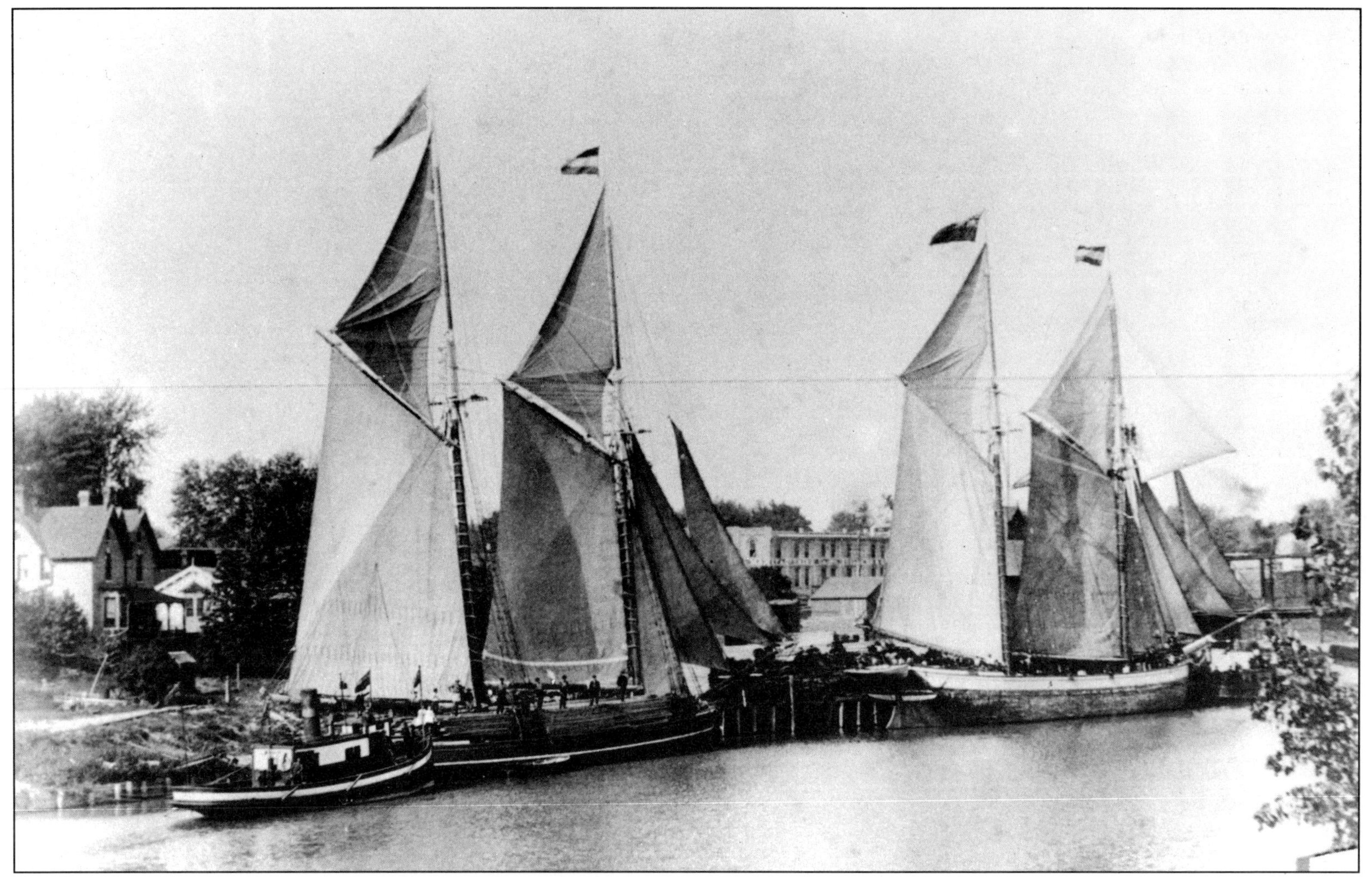

Marine Museum of Upper Canada, Toronto

White Oak **and** ***Azov*** **at Chatham, Ontario**

Chatham, Ontario, lies several kilometres up the Thames River from Lake St. Clair. The handiness of small schooners like the *White Oak* and *Azov* (see also pg. 40) in servicing such ports is apparent. While their sails are set in this picture, it would have been the tug astern of *White Oak* that would have pulled them up the long bends of the river.

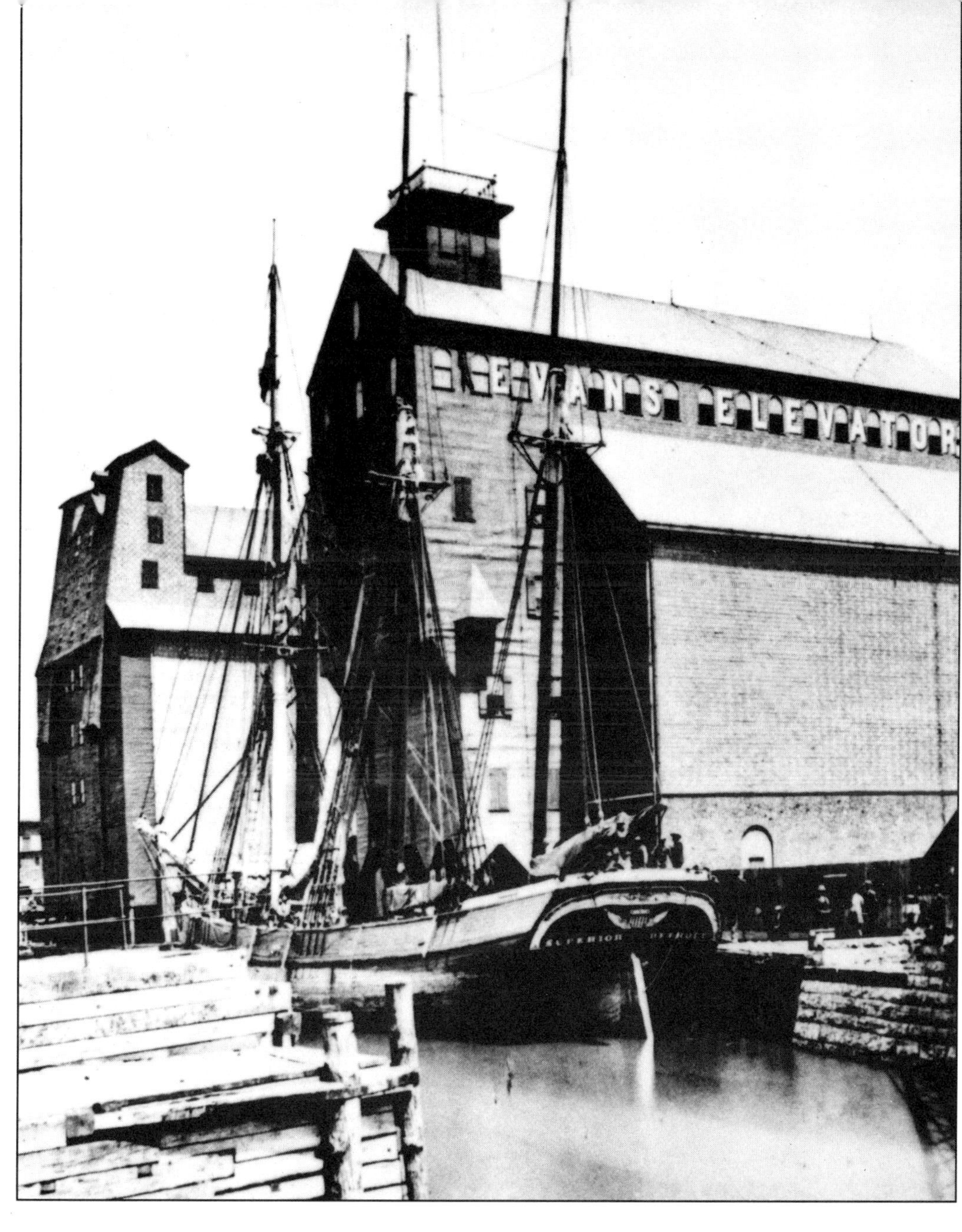

Superior

The *Superior* of Detroit is pictured beside the Evans Elevator on the Buffalo waterfront. It was probably unloading grain brought from the burgeoning farm states south and west of the Great Lakes. The grain was milled into flour or transferred to Erie Canal barges to be shipped east. The painted carving of a stars and stripes shield and flags decorating the transom is typical of the care and pride lavished on even working vessels in the age of wooden ships (see pp. 26 and 28).

Buffalo and Erie County Historical Society

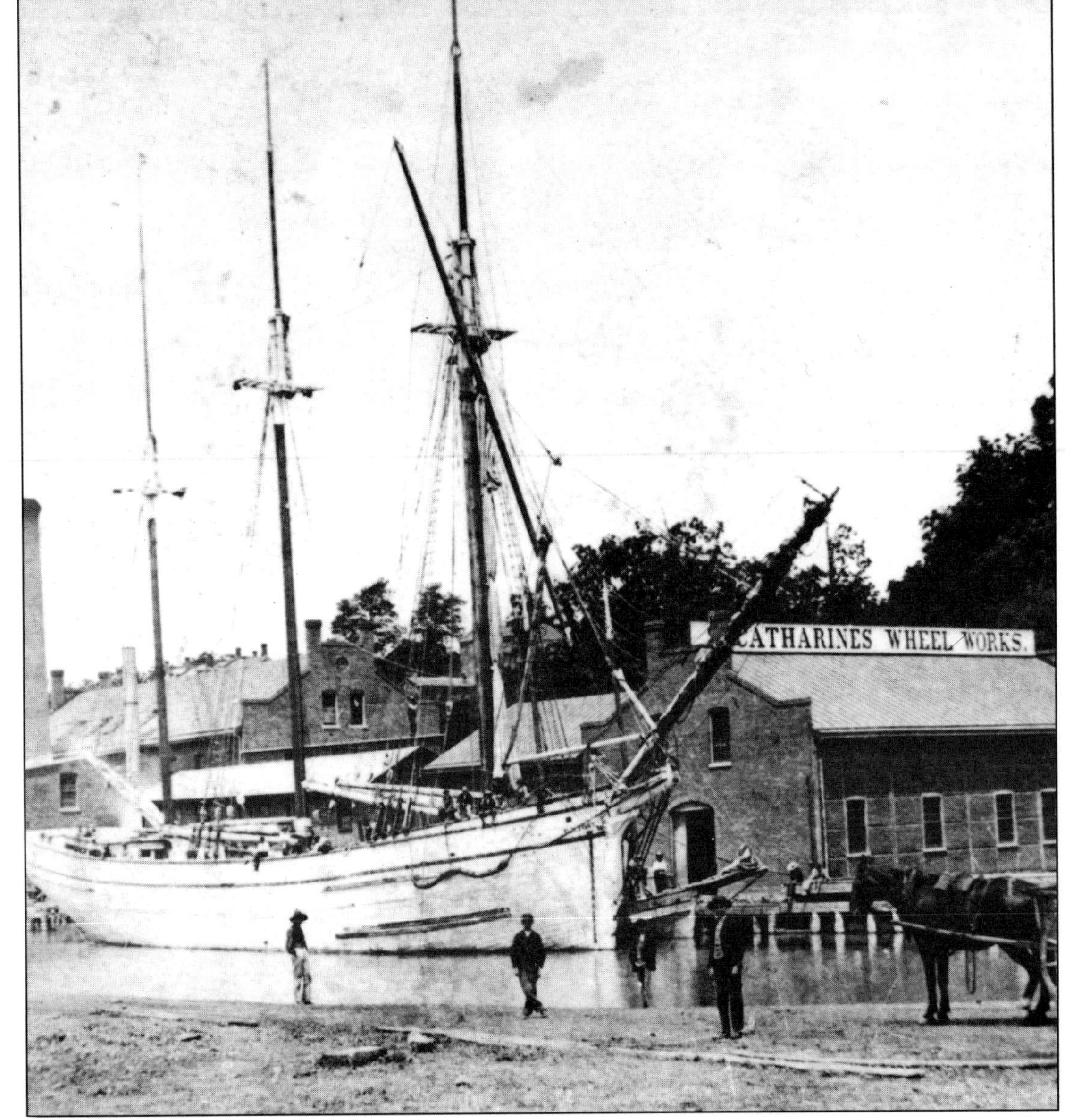

D. M. Foster

Many industries grew up in ports and along the canals to supply manufactured products to the steadily growing population in the Great Lakes basin. The *D. M. Foster* is seen at a wharf beside the St. Catharines Wheel Works on the Welland Canal in Ontario. The bowsprit and jib-boom have been raised to allow easier passage through the canal locks.

National Archives of Canada

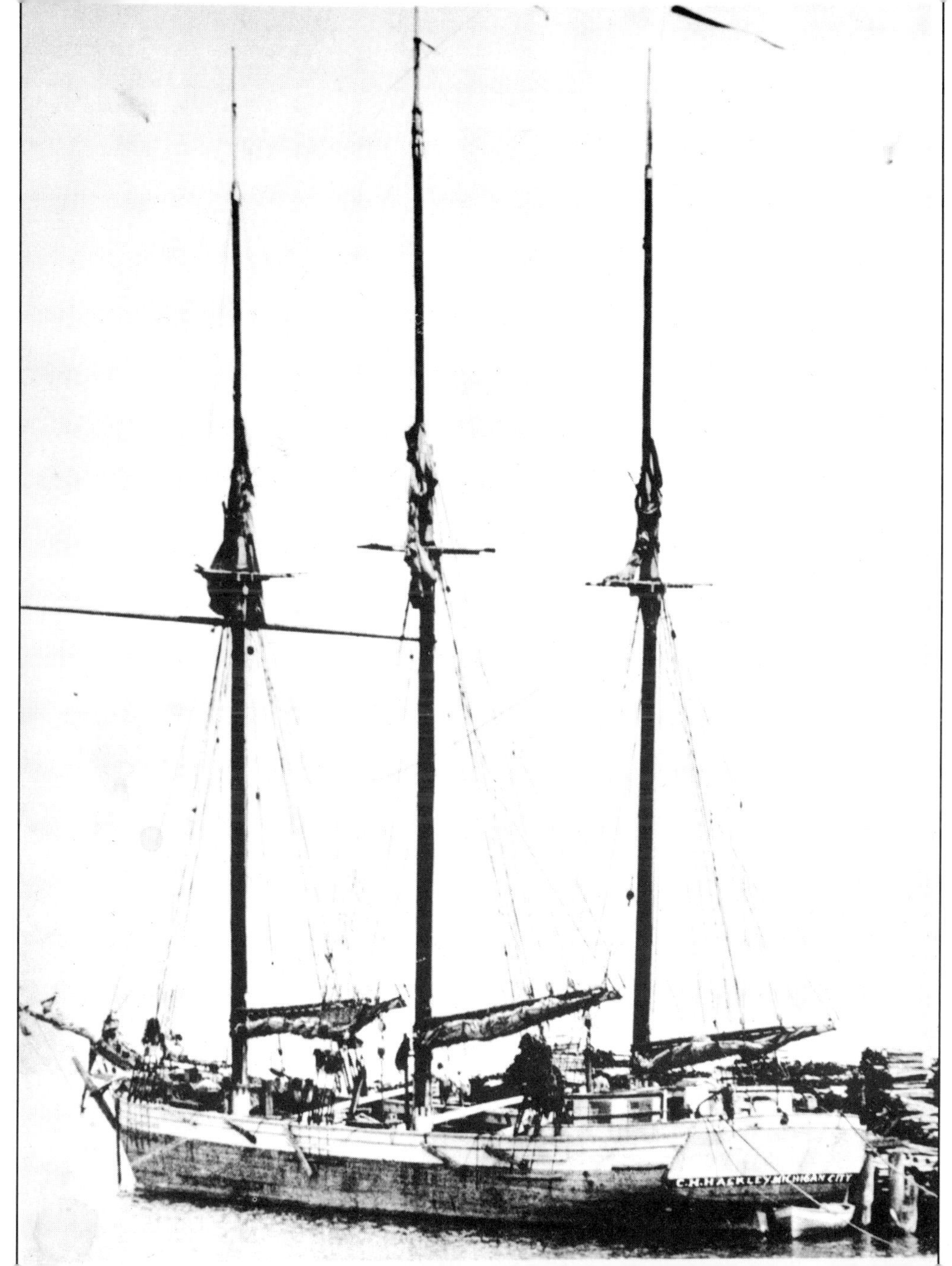

C. H. Hackley

In the nineteenth century, before the advent of gasoline or electric motors, most cargoes were loaded by hand. The schooner *C. H. Hackley*, named for the Michigan merchant who owned it, is seen loading lumber. Two ramps run over the side of the vessel down which the timber would be carried by longshoremen.

Great Lakes Historical Society, Vermilion, Ohio

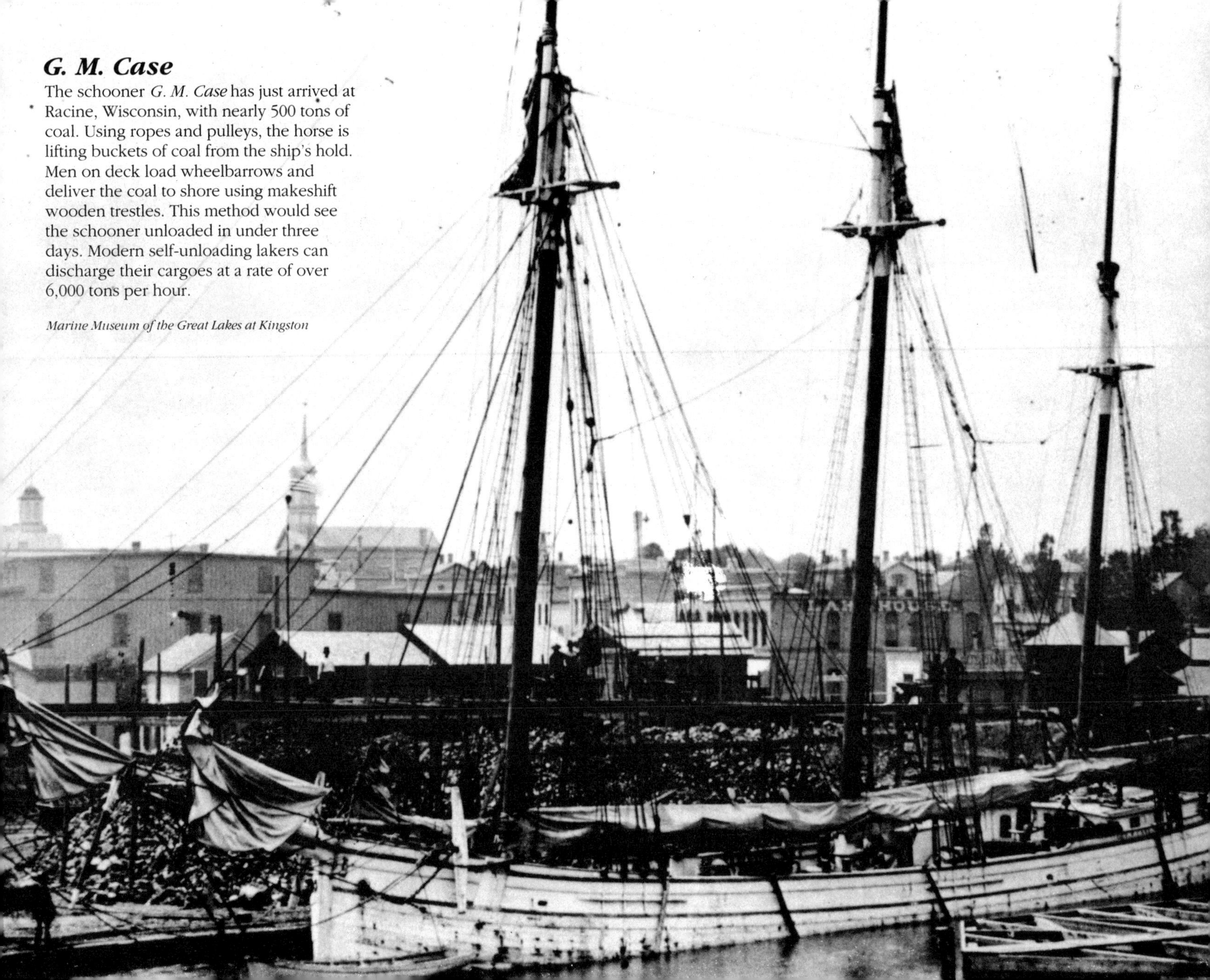

G. M. Case

The schooner *G. M. Case* has just arrived at Racine, Wisconsin, with nearly 500 tons of coal. Using ropes and pulleys, the horse is lifting buckets of coal from the ship's hold. Men on deck load wheelbarrows and deliver the coal to shore using makeshift wooden trestles. This method would see the schooner unloaded in under three days. Modern self-unloading lakers can discharge their cargoes at a rate of over 6,000 tons per hour.

Marine Museum of the Great Lakes at Kingston

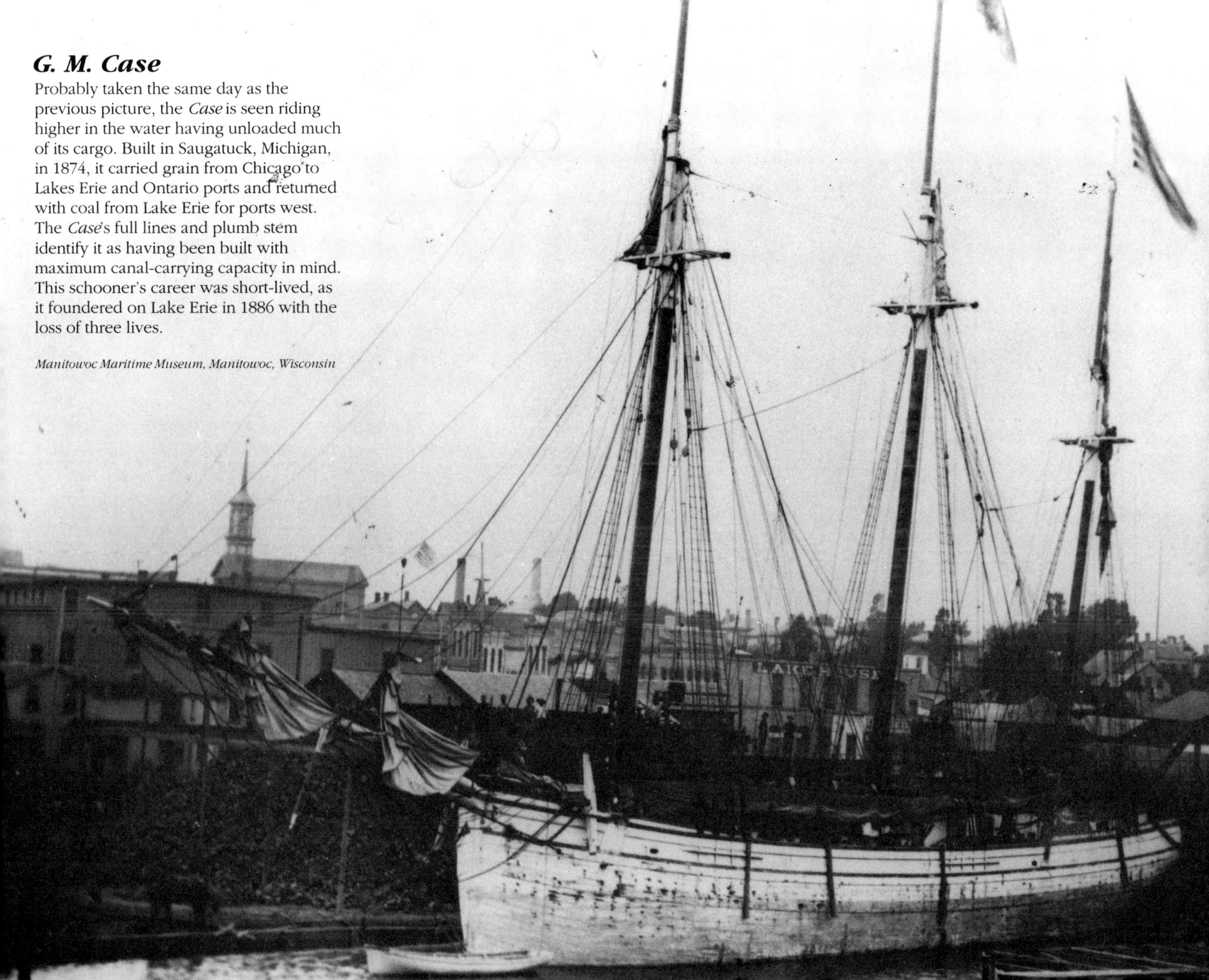

G. M. Case

Probably taken the same day as the previous picture, the *Case* is seen riding higher in the water having unloaded much of its cargo. Built in Saugatuck, Michigan, in 1874, it carried grain from Chicago to Lakes Erie and Ontario ports and returned with coal from Lake Erie for ports west. The *Case's* full lines and plumb stem identify it as having been built with maximum canal-carrying capacity in mind. This schooner's career was short-lived, as it foundered on Lake Erie in 1886 with the loss of three lives.

Manitowoc Maritime Museum, Manitowoc, Wisconsin

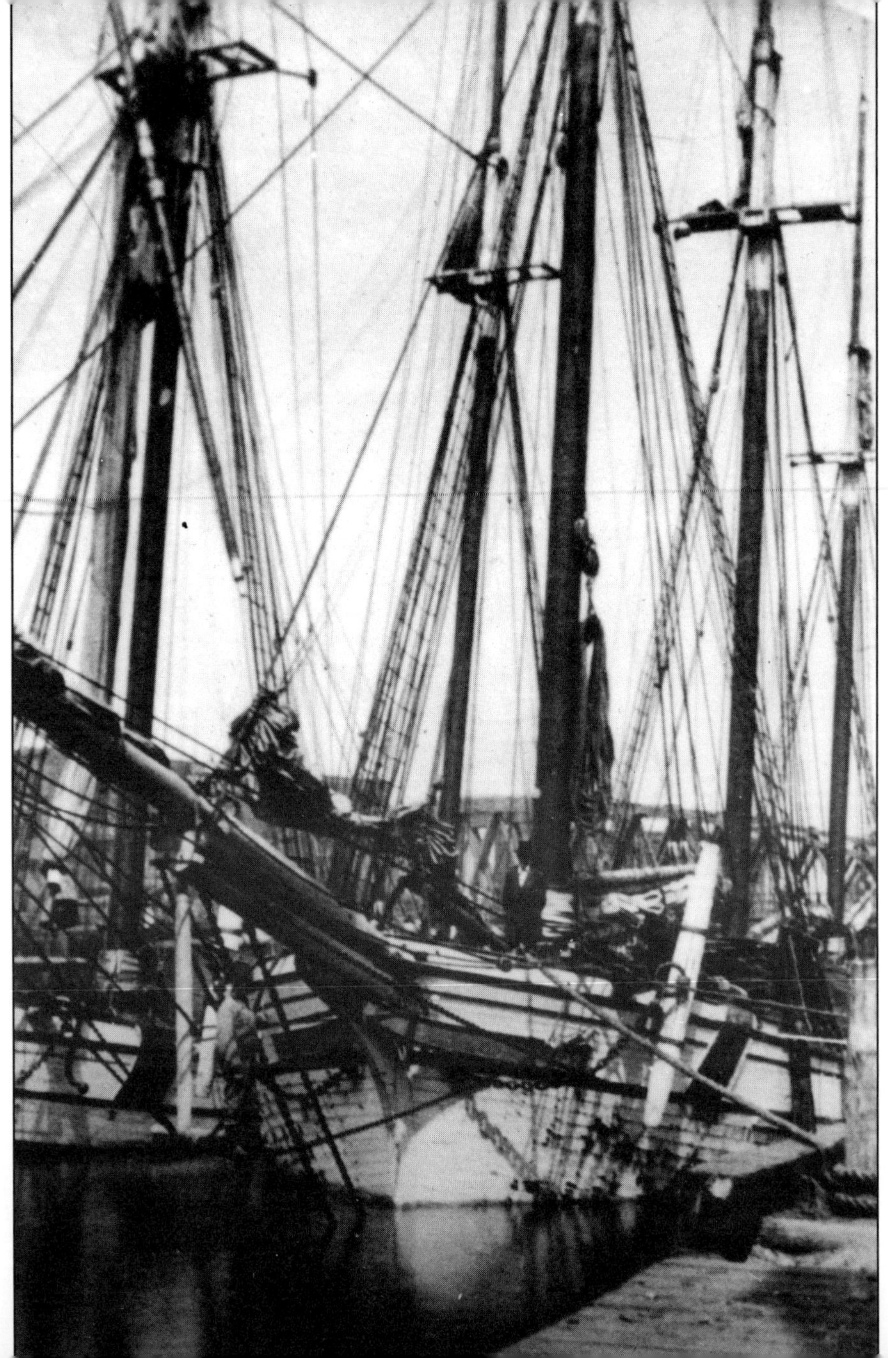

Schooners in Port

There was always work to be done around sailing vessels. The sailor on the left of this picture is examining the bow of the ship, while standing on one of the chains that secured the bowsprit and jib-boom. Some sense of the complexity of the rigging and the challenge for a sailor to "know the ropes" can be clearly grasped from this picture. Perhaps the man on the deck in the top hat is the owner.

Manitowoc Maritime Museum, Manitowoc, Wisconsin

LIFE ABOARD

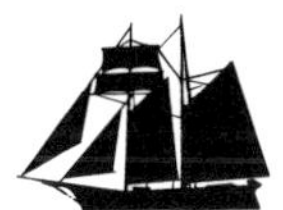

The crew of the typical Great Lakes schooner of the nineteenth and early twentieth century consisted of a captain (often the owner or part-owner), a mate, two or three crew and sometimes a cook. Often the cook was a woman, but with such small crews she was often expected to assist with numerous seafaring duties as well. Old records bear this out. We read of Minerva Ann McCrimmon, cook aboard the schooner *David Andrews*, who was at the wheel when the ill-fated vessel went ashore near Oswego, New York, in 1880, and Jane Hyatt of the schooner *Huron* who helped save the ship during a storm.

While some crews were stable—the same sailors from the same home harbour working the vessel year after year—one is struck by how often the records show crews with members all from different ports. Many of these seamen signed on for only a few trips at a time before moving on. It is interesting too, how often crews included members from both sides of the lakes. Both Canadian and American ports on the Great Lakes looked more toward the water for their contacts than inland to the rest of their own countries.

The crews lived in the cabin that was generally built into the after end of the vessel, but unlike their saltwater counterparts, they were not at sea for prolonged periods of time. A lakes voyage lasted for weeks rather than months. The winter freeze-up on the inland seas also meant that mariners could spend at least three or four months at home, although the economic reality often meant looking for other work in the off-season.

Life for Great Lakes schooner crews was not an easy one. There was always work to be done around sailing ships and much of that work was dangerous. While the fear of falling from the masthead tops or being injured by machinery was ever-present, the worst prospect was the weather, particularly the storms the lakes so often produced at the end of the shipping season. In one year alone, 1883, 672 mariners lost their lives in the wrecks of forty vessels on the lakes. Mariners Church, a shrine that most of the sailors passed often on the Detroit waterfront, is said to have tolled its bell once for each person lost on the lakes over the years.

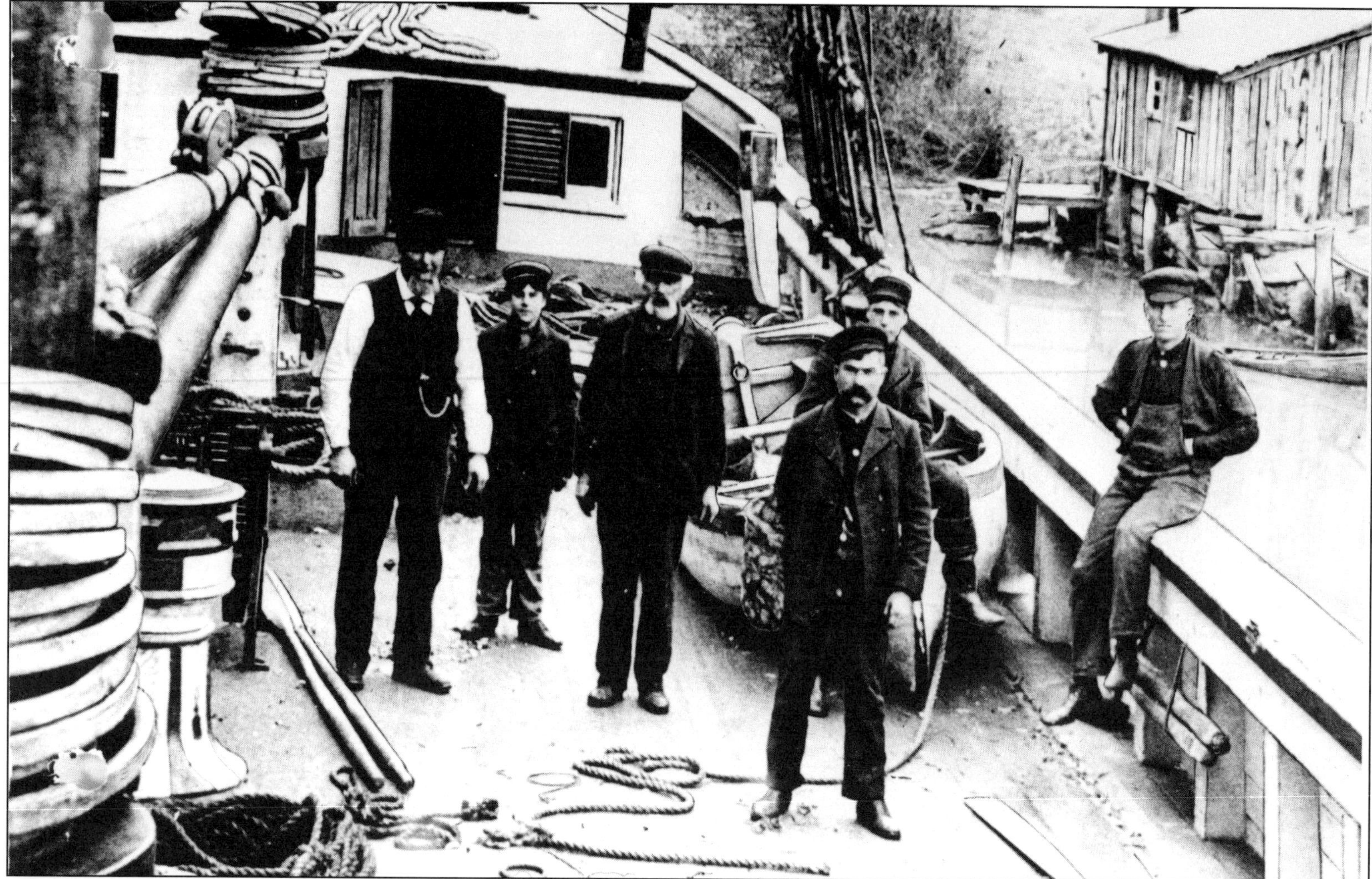

Crew of the Schooner *Azov*

Marine Museum of Upper Canada, Toronto

Some appreciation of the work environment of the schooner crew can be garnered from this picture aboard the Azov. The after cabin where the crew lived is visible behind the posing figures. At the left, just behind the mast, is the capstan which was turned by hand to haul lines when raising sails and carrying out other manoeuvres. The bulwarks along the right side kept the crew from being swept over the side in rough weather. It appears that two of the crew are advanced in age, two are quite young and two are enjoying their middle years. Such an age spread was probably not uncommon.

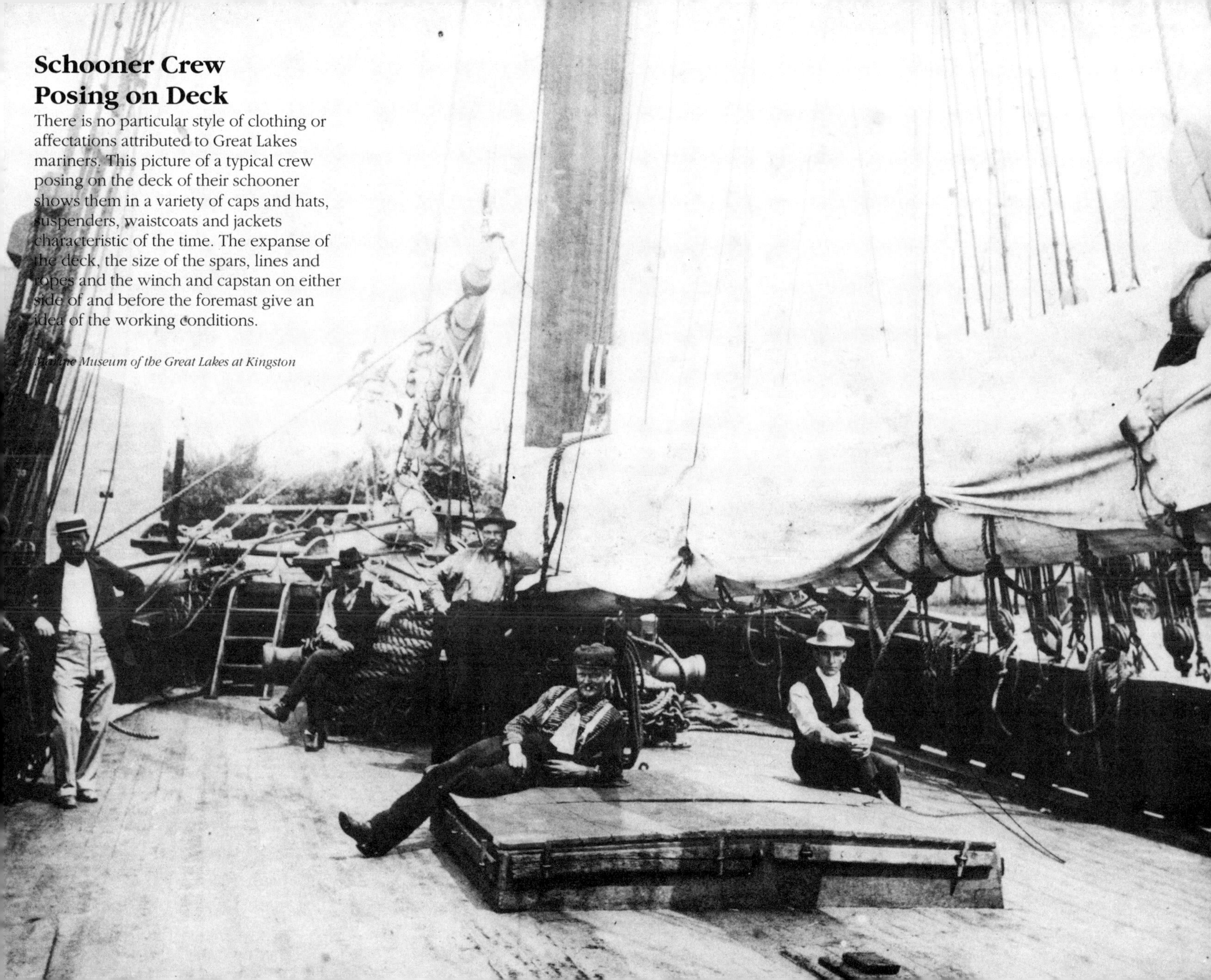

Schooner Crew Posing on Deck

There is no particular style of clothing or affectations attributed to Great Lakes mariners. This picture of a typical crew posing on the deck of their schooner shows them in a variety of caps and hats, suspenders, waistcoats and jackets characteristic of the time. The expanse of the deck, the size of the spars, lines and ropes and the winch and capstan on either side of and before the foremast give an idea of the working conditions.

Marine Museum of the Great Lakes at Kingston

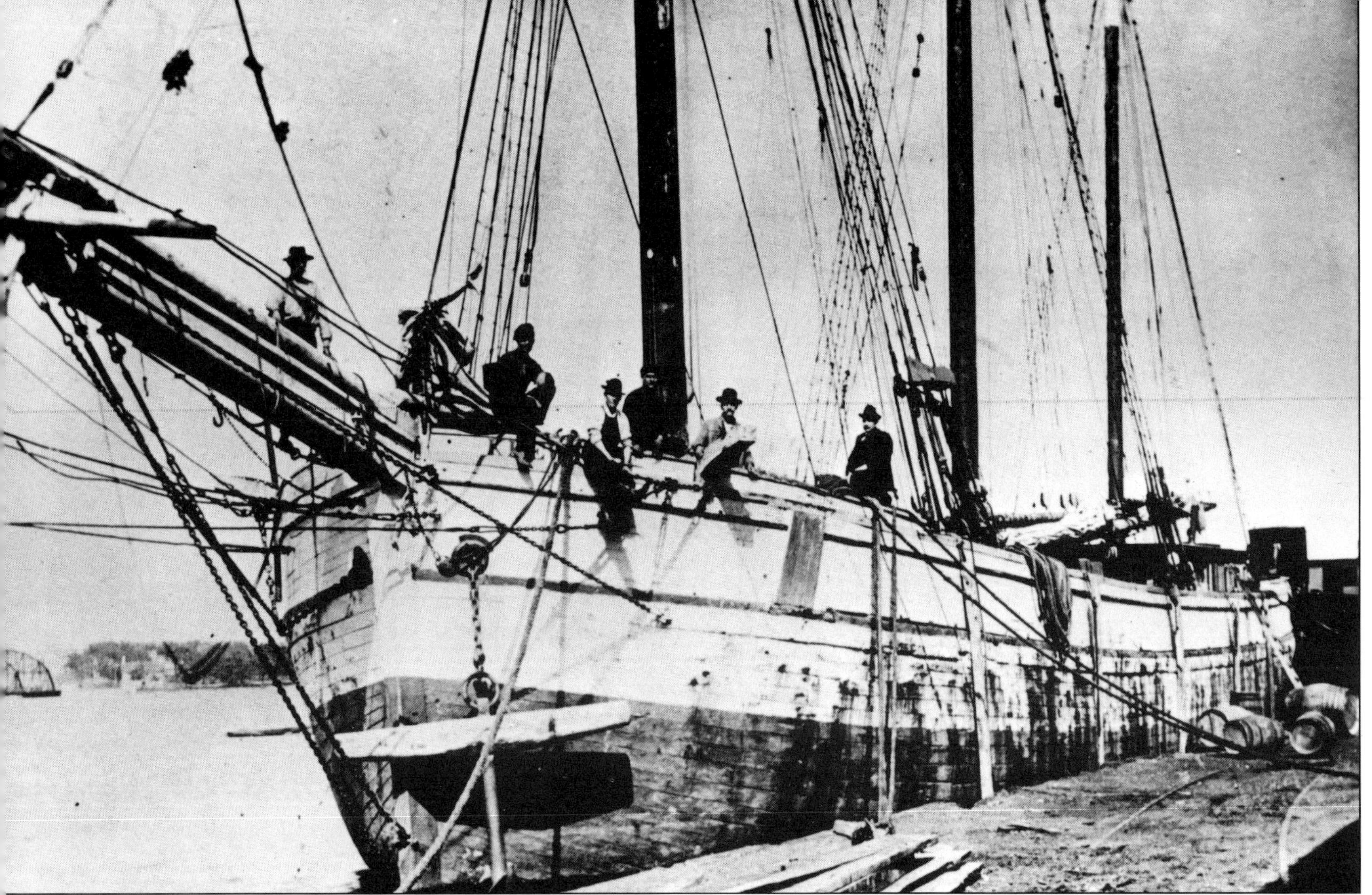

Marine Museum of Upper Canada, Toronto

Keewatin

The typical six-man schooner crew can be seen here in proportion to the vessel they sailed. On the wharf beside the *Keewatin* are the rails that brought railcars and unloading cranes to the side of the ship. As well, there are barrels of the type in which so much cargo was transported during the nineteenth century.

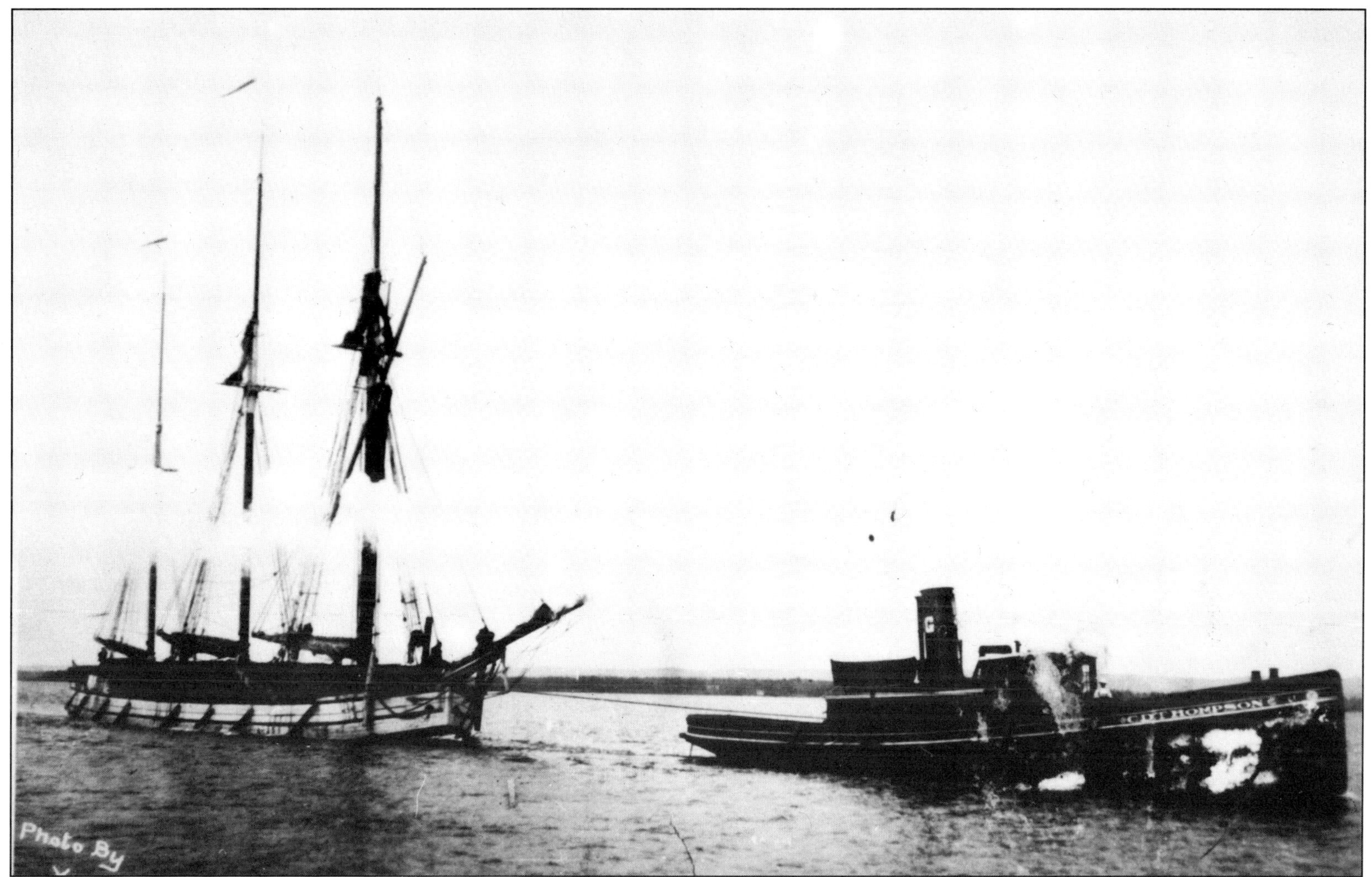

Great Lakes Historical Society, Vermilion, Ohio

Butcher Boy

The name of this vessel, which was built in De Pere, Wisconsin, in 1868, exemplifies one of the nomenclature patterns of lake ships. The vessel was named after a person, Mark Robbins of Chicago, Illinois, but not directly. Mark was young and his father was a principal in the meat packing industry. Hence, *Butcher Boy.* Young Mark lost his life in the great Chicago fire of 1871, which also damaged the schooner. The vessel, however, survived until 1917.

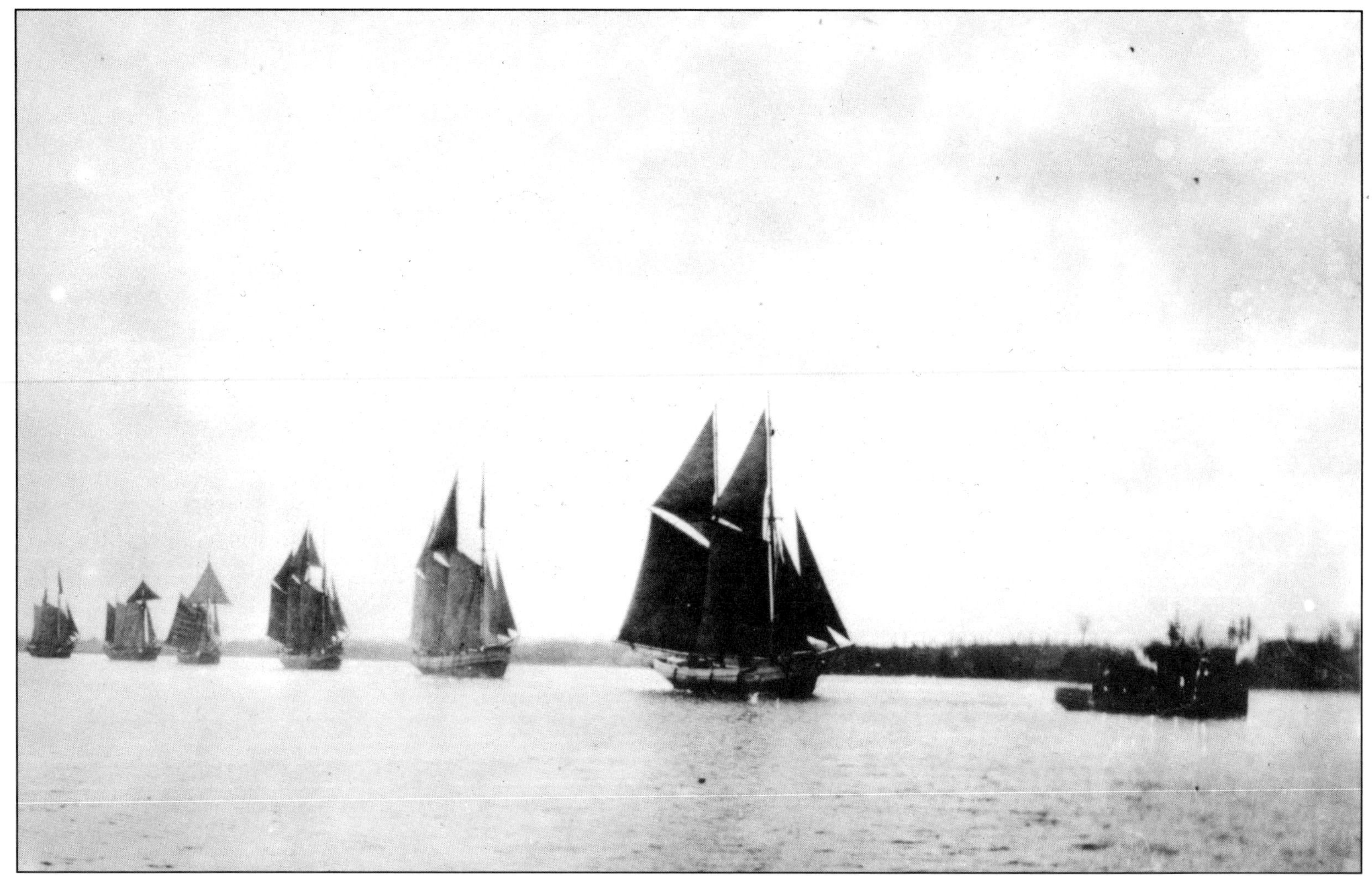

A Long Tow

Great Lakes Historical Society, Vermilion, Ohio

Butcher Boy, from the previous page, is the second to last vessel in this picture of a long tow behind the tug *Goldsmith*. The others are *Fearless*, *York State*, *Ever Fuller*, *Porter* and *George Steele*. Increasingly throughout the nineteeth century steam tugs were employed to assist sailing schooners. Particularly on the Detroit and St. Clair Rivers, where this picture was probably taken, schooners were pulled in long lines. This saved them the time that would have been required to manoeuvre under sail power alone and reduced the chances of costly accidents.

TOWING AND CANALING

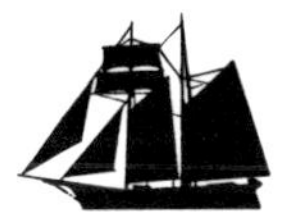

Before the age of steam, schooners had to work their way in and out of harbour under their own sail power. This was often dangerous and usually time consuming.

The Erie Canal was navigated only by specially constructed barges, but when the Welland Canal was planned in the 1820s, it was to be a ship canal capable of passage by many sailing vessels of the day. The St. Lawrence and the Soo canals were similarly conceived. Once in these waterways, vessels had to be towed. Paths were constructed which ran continuously along one side of the canal to allow the passage of teams of draught animals employed in towing. For larger vessels several teams of horses and sometimes oxen were used. The latter were referred to as "horned wind."

As soon as steam power was perfected and applied to small work boats, however, the task of pulling vessels in confined areas and through canals passed largely to tugs. The Welland was the longest of the canals and had by far the greatest lift. Draught animals continued late into the nineteenth century to be used to tow ships through the areas where most of the locks were located close to Lake Ontario. Even the Third Welland Canal constructed in the 1880s was provided with a tow path. But through the longest reaches of the canal where there were no locks tugs were used, and eventually the teams of horses gave way completely to steam tugs (see pp. 46, 48 and 49).

The trend to steam towing grew in areas other than harbours and canals. Sometimes long lines of schooners were towed together through the Detroit and St. Clair Rivers (see previous page).

As time went on more and more schooners were cut down to barges by having large portions of their sailing rigs removed. These were then towed from port to port as a matter of regular practice.

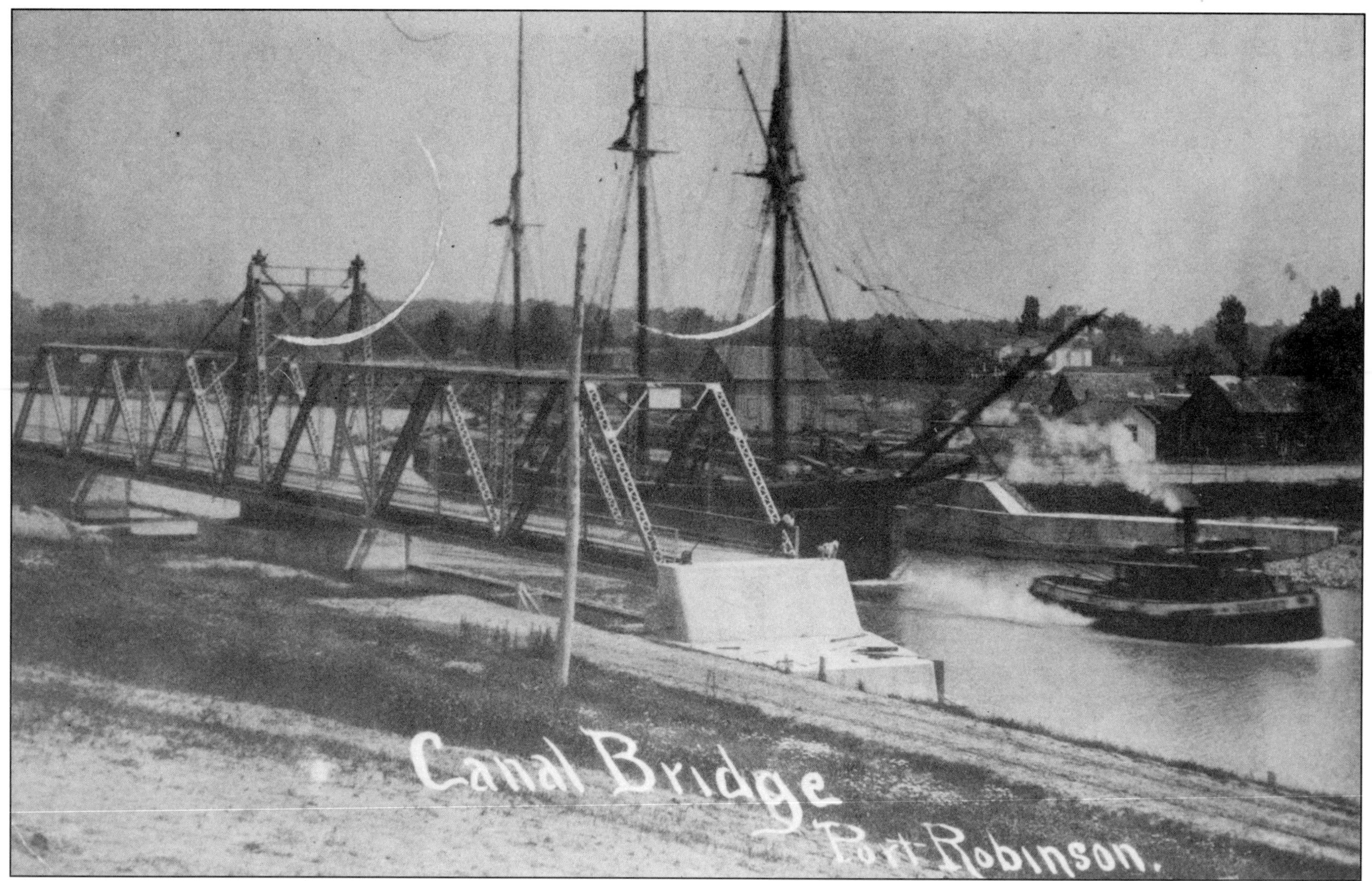

John Burtniak

Port Robinson

A common sight on the Welland Canal prior to 1920, a schooner passes through the draw of the swing bridge, upbound at Port Robinson. Tugs carried on a thriving business towing sailing ships and barges through the canals. This schooner has its bowsprit raised up and the raffee boom braced. These were standard practices to keep the spars from becoming entangled with other vessels when the ship was moving in the close confines of a canal.

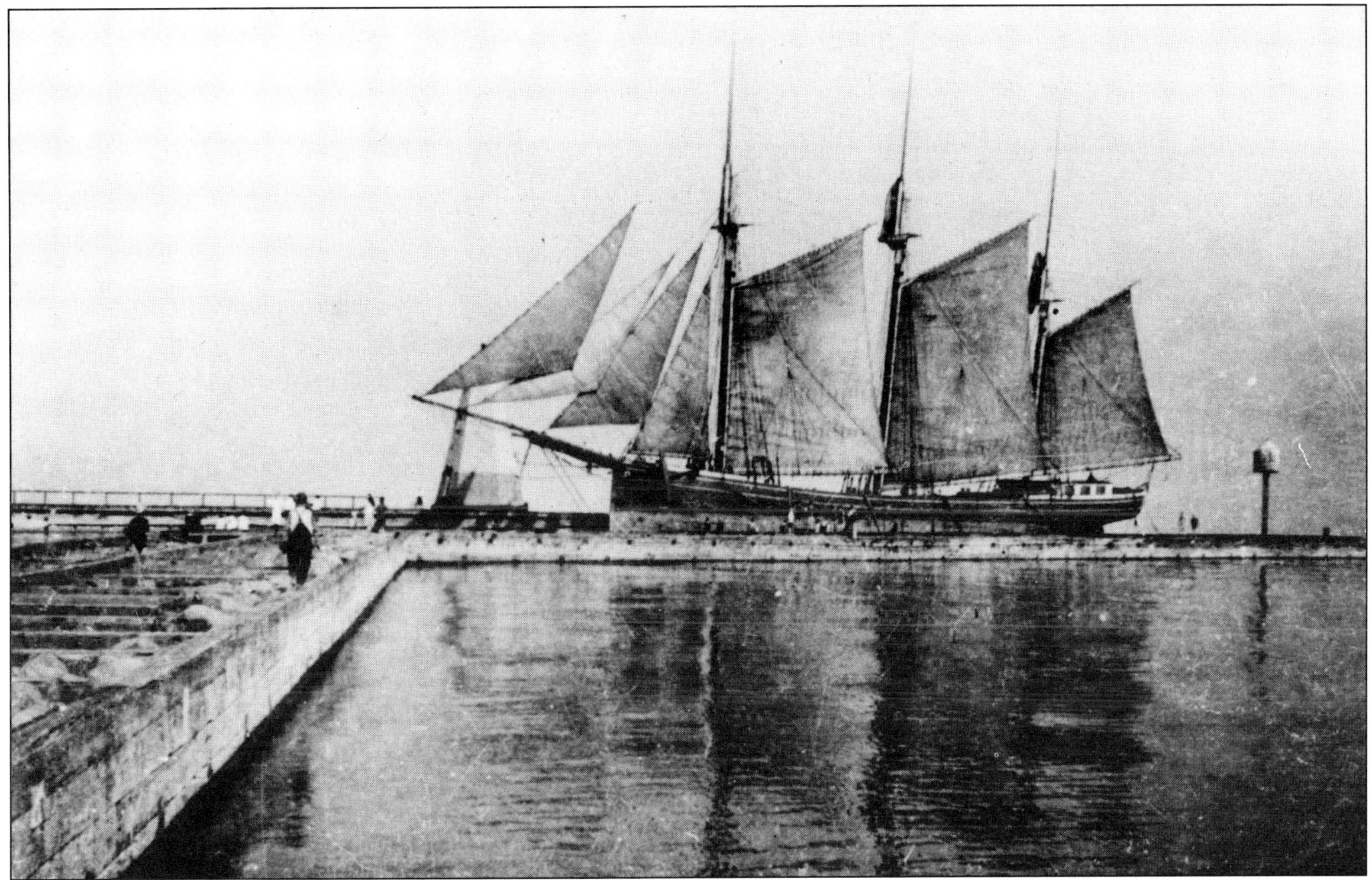

National Archives of Canada

Arthur

The three-masted schooner *Arthur* was built in Manitowoc, Wisconsin, in 1873. Ten years later, with the opening of the Third Welland Canal, the *Arthur* was competing with steamboats 30 metres (100 feet) longer and capable of carrying twice as much cargo. Approaching the piers at Toronto, the schooner is waiting for a tug to tow it to its assigned berth for loading.

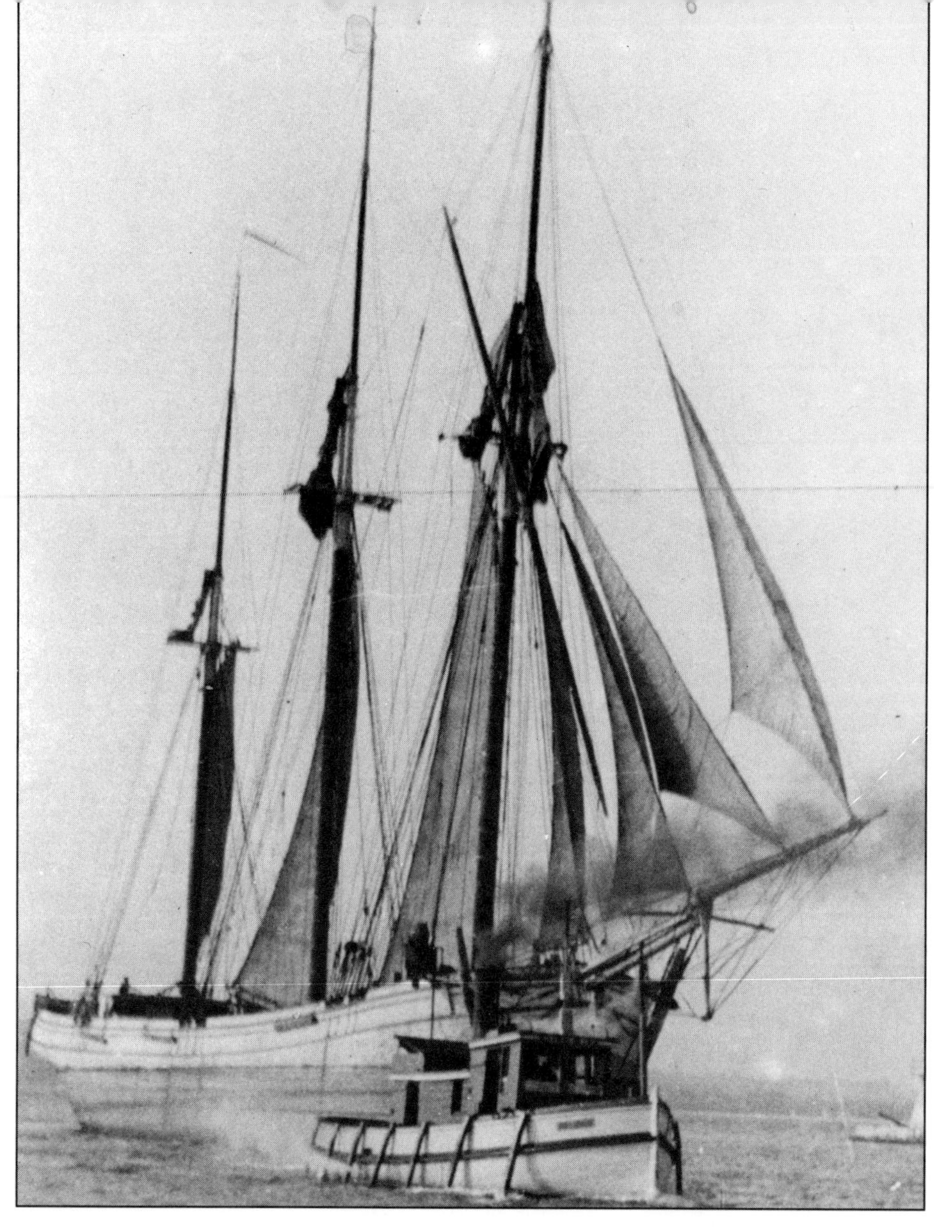

Arthur

The timely arrival of a tug could mean the difference of several important hours and sometimes days in a schooner's voyage. The canals were closed on Sundays and traffic congestion could extend the ship's transit time to a day and a half. Modern vessels can transit the present-day Welland Canal in less than twelve hours.

Marine Museum of the Great Lakes at Kingston

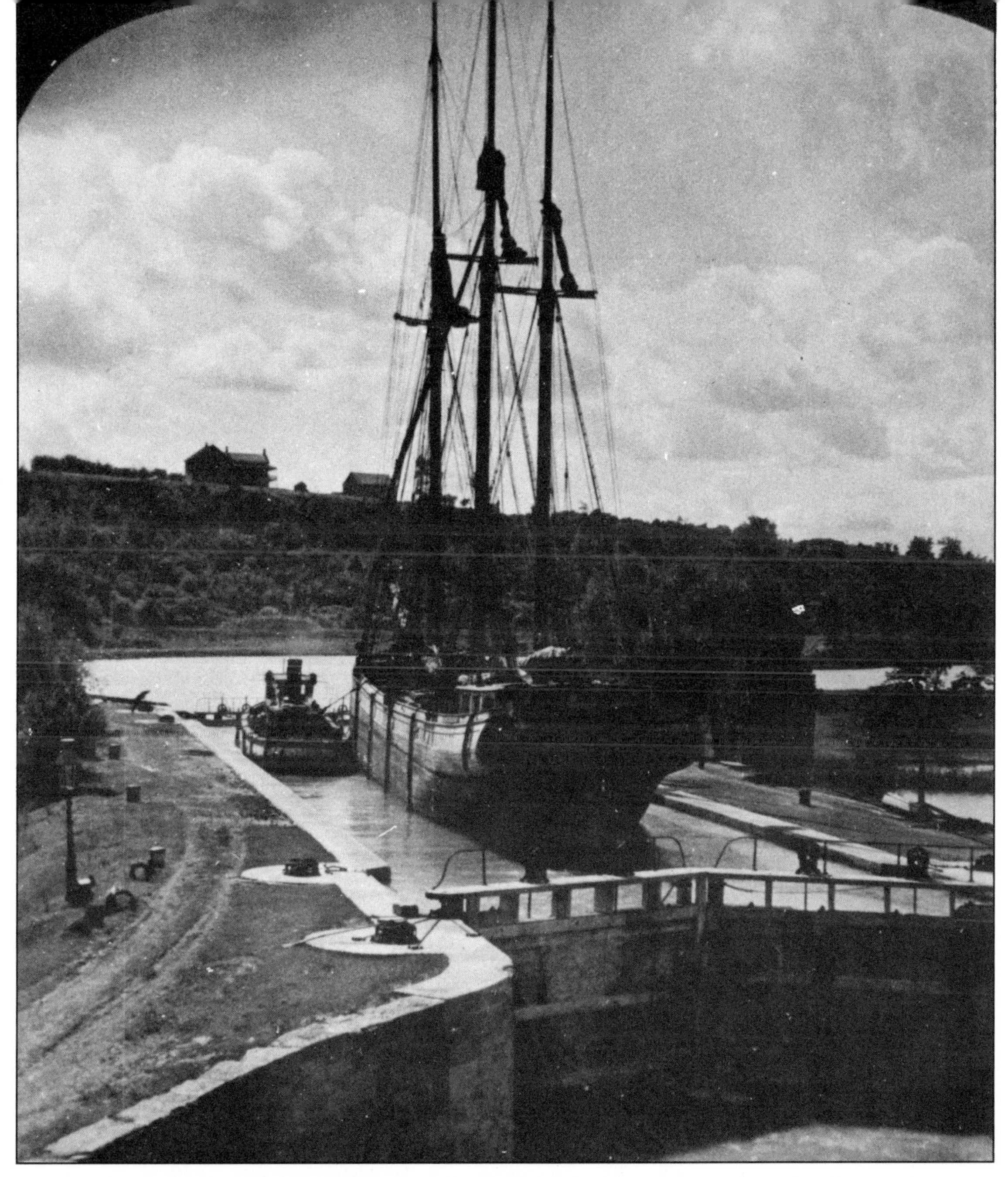

Arthur

Welland Canal locks, enlarged in the 1880s, show the carrying capacity disadvantage of the schooner toward the turn of the century. Tugs accompanied schooners through canals, picking up their towlines upon entering and escorting them safely into the lake after clearing the final lock. In 1880 tug services for a canal transit totalled $14. The *Arthur* was sold for off-lakes service in 1918 and made three round-trips between Halifax, Nova Scotia, and South Africa.

Public Archives of Ontario

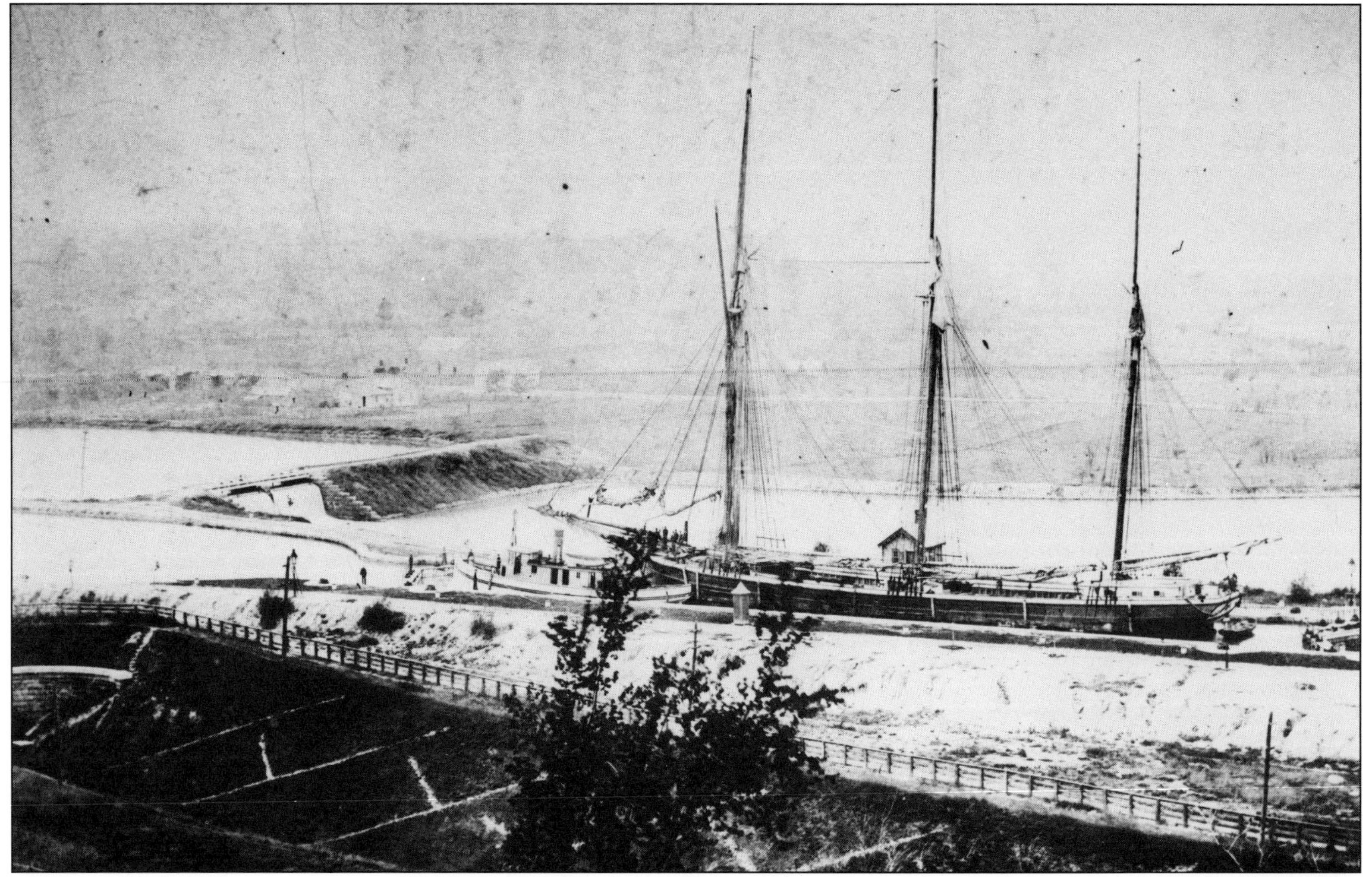

St. Catharines Historical Museum

Mountain Locks

The slowest part of the journey through the Great Lakes was often the so-called mountain locks of the Welland Canal. At the point where the canal ascended the Niagara Escarpment in its circumvention of the great falls of Niagara, there were a dozen locks each within sight of the next. In this picture a large schooner is negotiating the canal at the place where the Grand Trunk Railway line ran under the waterway through a tunnel.

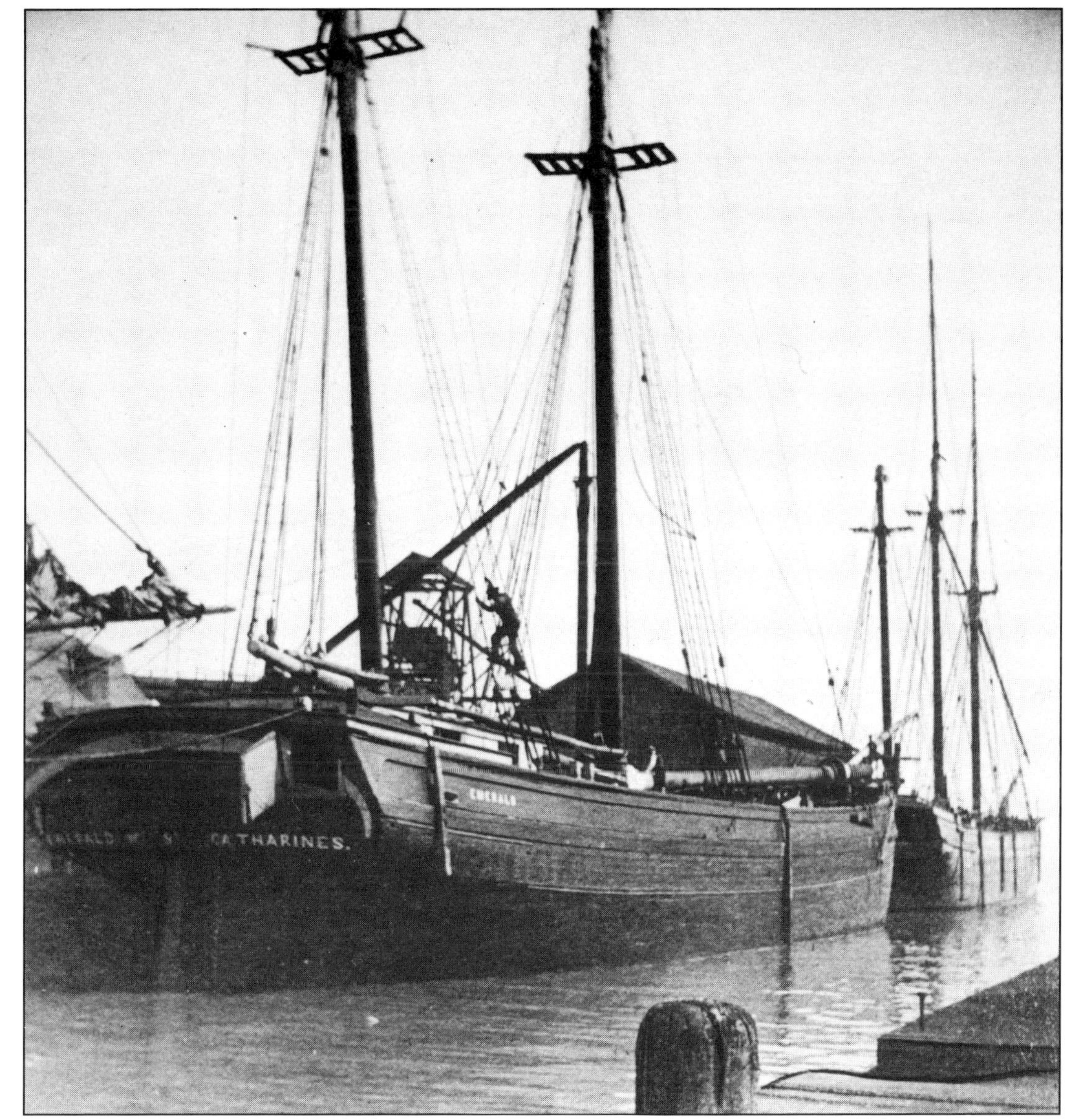

Emerald

The schooner *Emerald* built in 1872 at Port Colborne, Ontario, featured loading ports in its bow and stern. This allowed large timbers to be hauled through the ports into the hold (see also pg. 26). The *Emerald* (see pg. 58) disappeared without a trace in 1911 while en route from Charlotte, New York, to Cobourg, Ontario, with a load of coal. The vessel in front of *Emerald* is the *St. Louis*, which was built by Louis Shickluna in St. Catharines, Ontario, in 1877.

Marine Museum of Upper Canada, Toronto

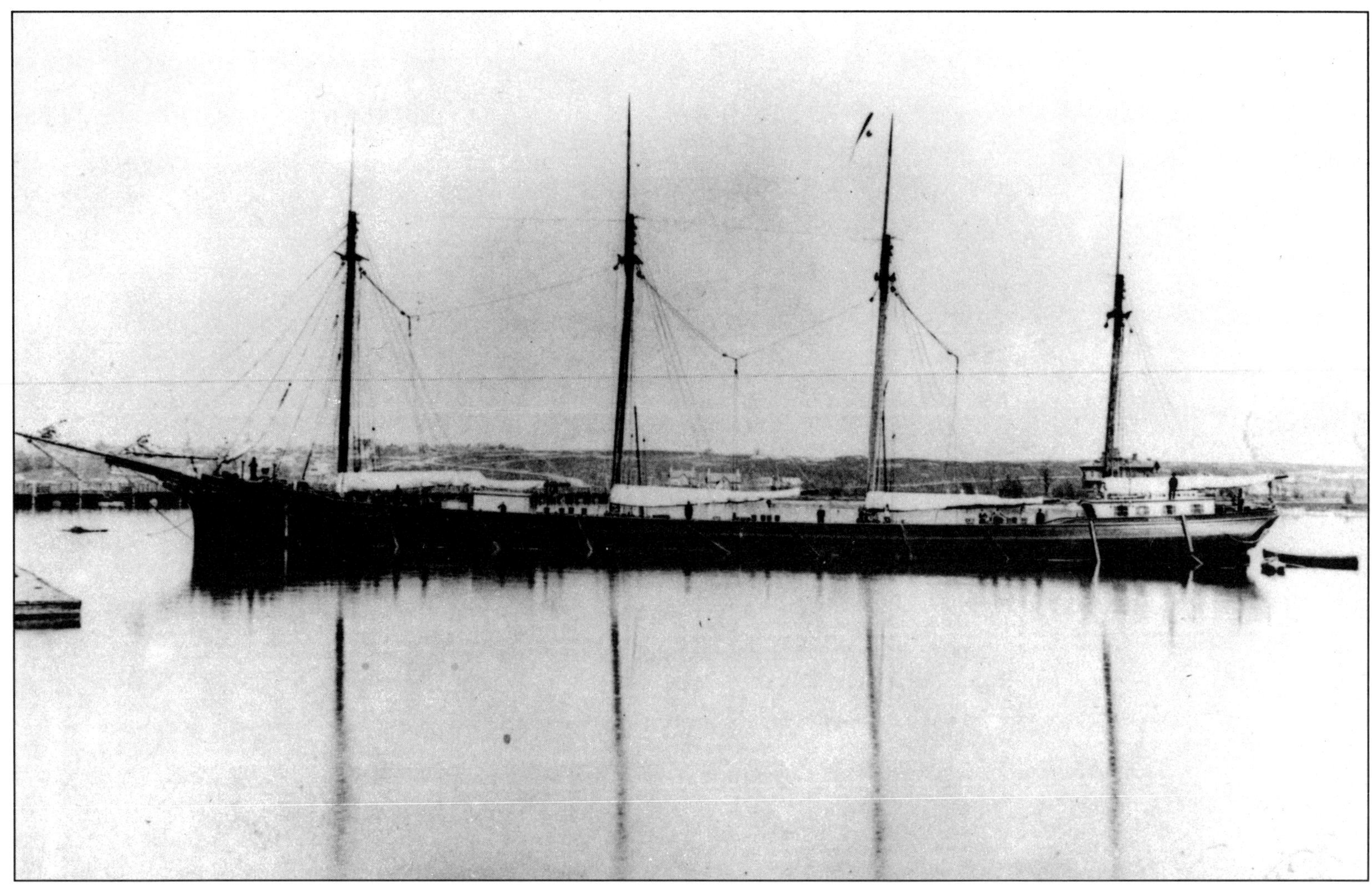

National Archives of Canada

Minnedosa

The four-masted schooner *Minnedosa* was the largest Canadian schooner built on the Great Lakes. Although rigged as a schooner, the *Minnedosa* only operated as a schooner-barge under tow. Measuring 76 metres (250 feet) in length, it sank in Lake Michigan in 1905 while carrying a cargo of 75,000 bushels of wheat from the west to Kingston, Ontario. By comparison, today's modern Seaway-sized lakers carry over a million bushels of grain.

TRANSITION TO STEAM

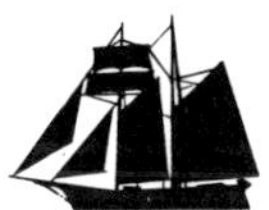

The transition from sail to steam power on the Great Lakes did not happen quickly nor was the changeover automatic. It has often been noted with irony that the zenith of the saltwater sailing age, the advent of the great tea clippers, occurred after the introduction of steam. To some extent the same phenomenon occurred on the Great Lakes.

There continued to be many advantages to sail and to schooners in particular. Wind was free at a time when coal was still relatively expensive. And schooners, which required a crew of only four, five or six, were cost effective. This was especially true in the early days of steam when very skilled engineers were required to keep the complex and temperamental machinery operating.

Add to this the fact that steam tugs could be used in the most difficult areas for schooners, to tow them in and out of harbour and through canals, and it can be seen that many reasons persuaded investors and merchants to continue building wooden sailing schooners.

Around the turn of the century, however, the tide began to turn irreconcilably. Steam engines were more reliable and men to run them were more available. Iron and steel were taking over from wood as the prime building material for ships, and the size of vessels economically running on the lakes was increasing dramatically.

Few new schooners were built after the 1880s. Many of those afloat were modified. Some were cut down to barges by having all or part of their sailing rigs removed. These were then towed by tugs or as consorts behind cargo steamers. Others were converted by having engines, boilers and wheelhouses installed. The last remaining schooners on the Great Lakes operated until the 1920s. Some were sold into the saltwater trade to be lost on tropical reefs and some were burned as spectacles. By the 1950s virtually none of the hundreds of proud Great Lakes schooners was left.

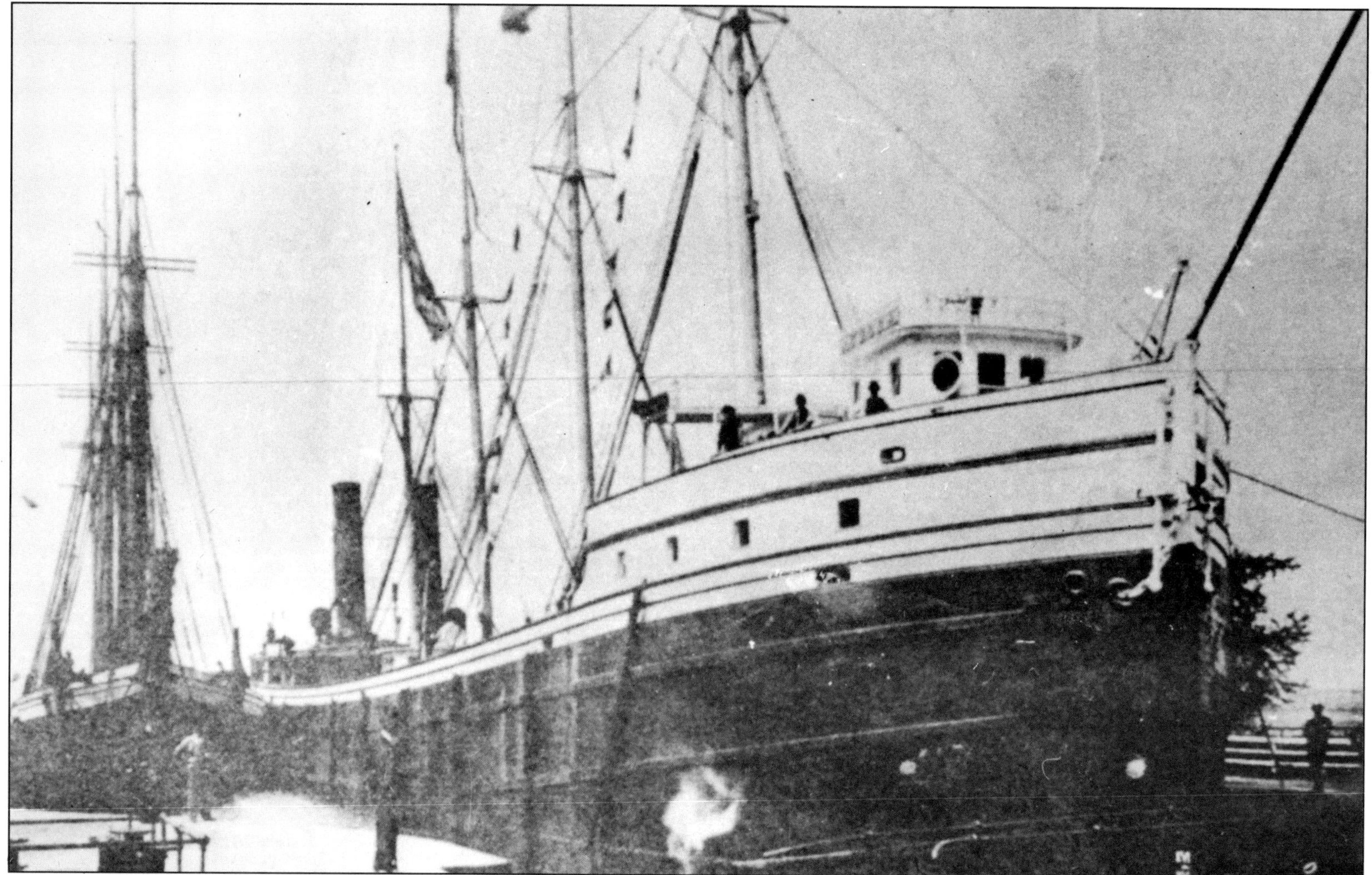

David Dows—steam barge

Institute for Great Lakes Research, Bowling Green State University

As with many schooners built in the final phase of sail power on the Great Lakes, the *David Dows* was modified after only a few years of work as a schooner. It had boilers, engines and a wheelhouse installed, but kept its masts. It was only eight years old in 1889 when it was lost in a November gale off Whiting, Indiana, in Lake Michigan.

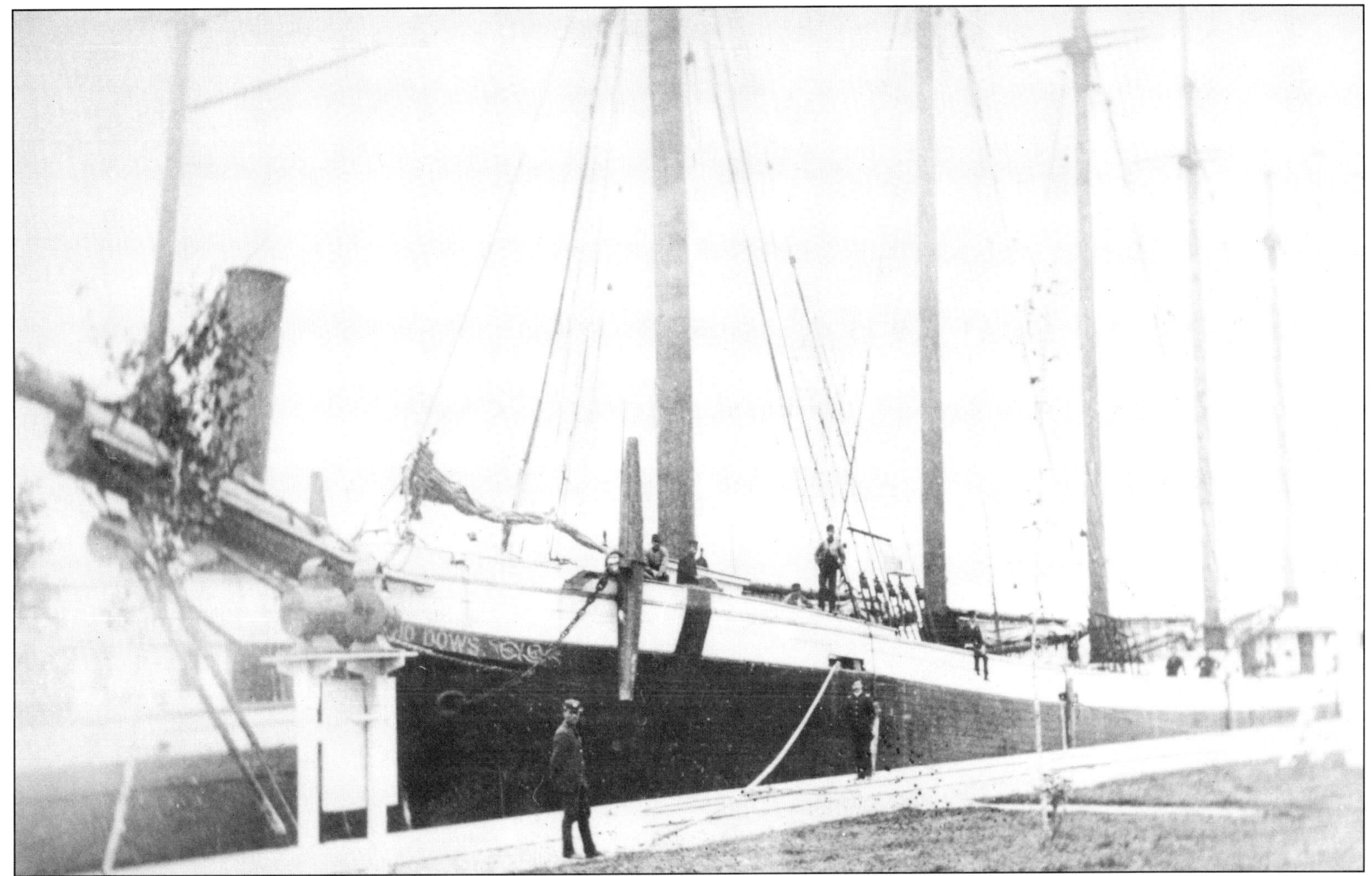

Institute for Great Lakes Research, Bowling Green State University

David Dows—schooner

The *David Dows* represents, in some ways, the height of development on the Great Lakes schooner. Built in 1881 on the banks of the Maumee River near Toledo, Ohio, it was 85 metres (278 feet) long and had five masts each over 49 metres (160 feet) tall. The *Dows* was not only the largest schooner ever built on the Great Lakes, but was reported to be a capable sailor as well, once going from one end of Lake Erie to the other in eighteen hours. That was faster than any steamer of the day.

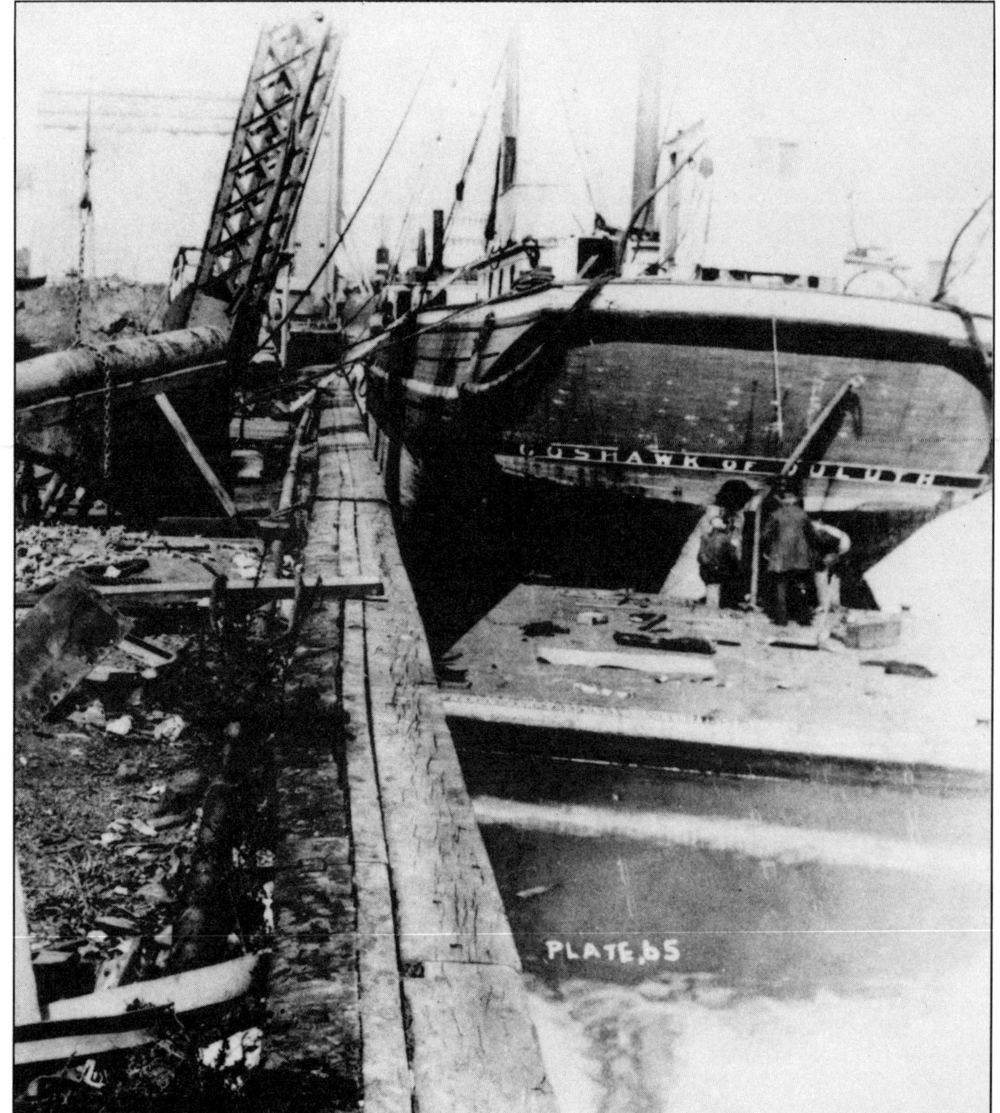

Goshawk

Maintenance of the old wooden schooner fleet became more and more of a problem after the turn of the century. Fewer craftsmen trained in the traditional skills were available and the shipbuilding and dry-dock facilities had been largely converted to service metal ships. Here we see the *Goshawk* of Duluth undergoing makeshift repairs to the rudder in Buffalo harbour.

Buffalo and Erie County Historical Society

Our Son →

This vessel was built at Black River, Ohio, in 1875 and was named in memory of the owner's son, who was drowned while the schooner was under construction. It was a smart sailor in its day, carrying more square-rigged sails on the foremast than most schooners. The boat that can clearly be seen in this picture slung over the transom was a characteristic feature of Great Lakes schooners. The *Our Son* was one of the very last sailing ships to operate on the lakes. It was finally lost in Lake Michigan in September 1930 after the crew were rescued by a steamer. The captain at the time was seventy years old.

From the Collection of Mr. Al Sagon - King, Thorold, Ontario

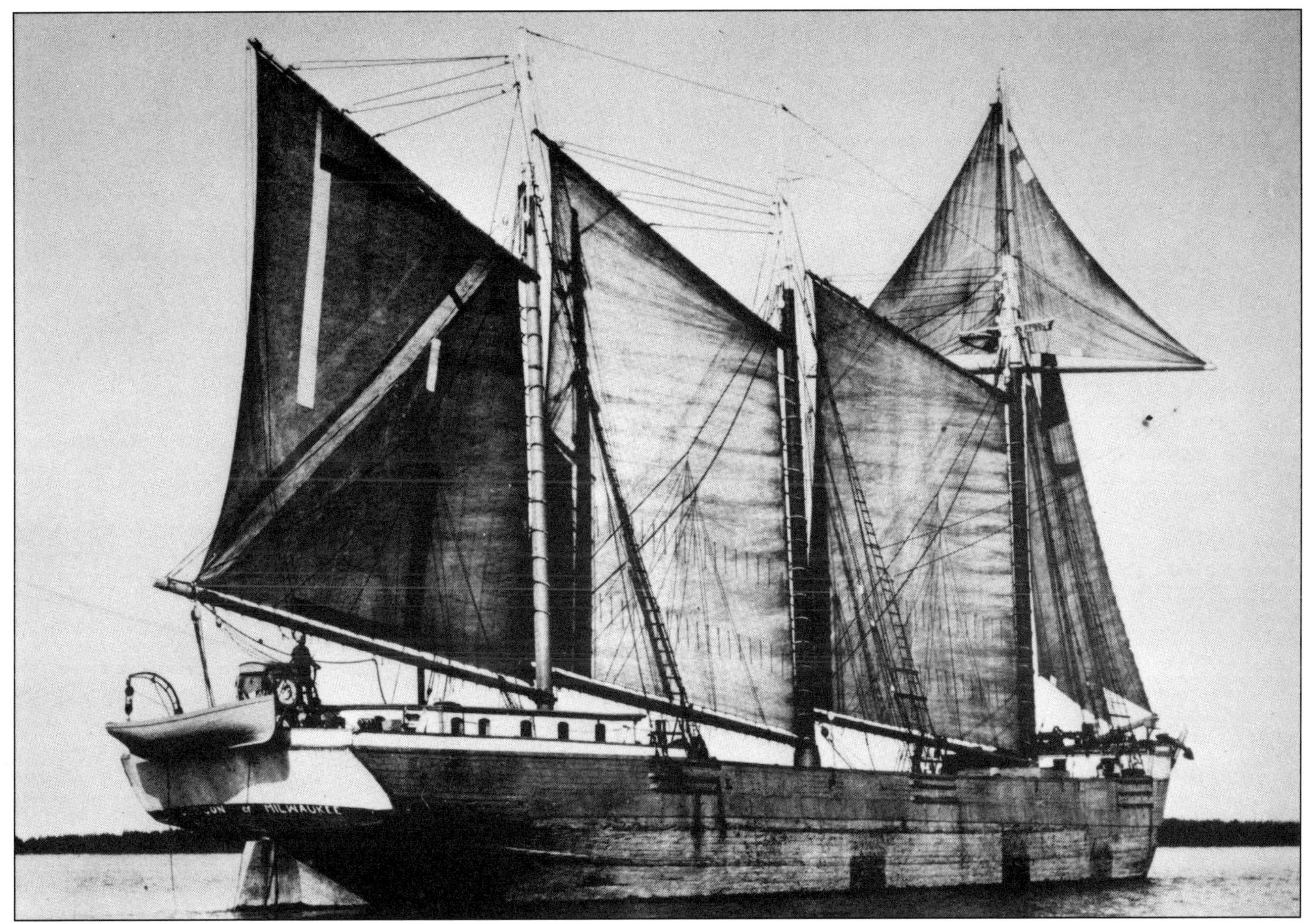
MILWAUKEE

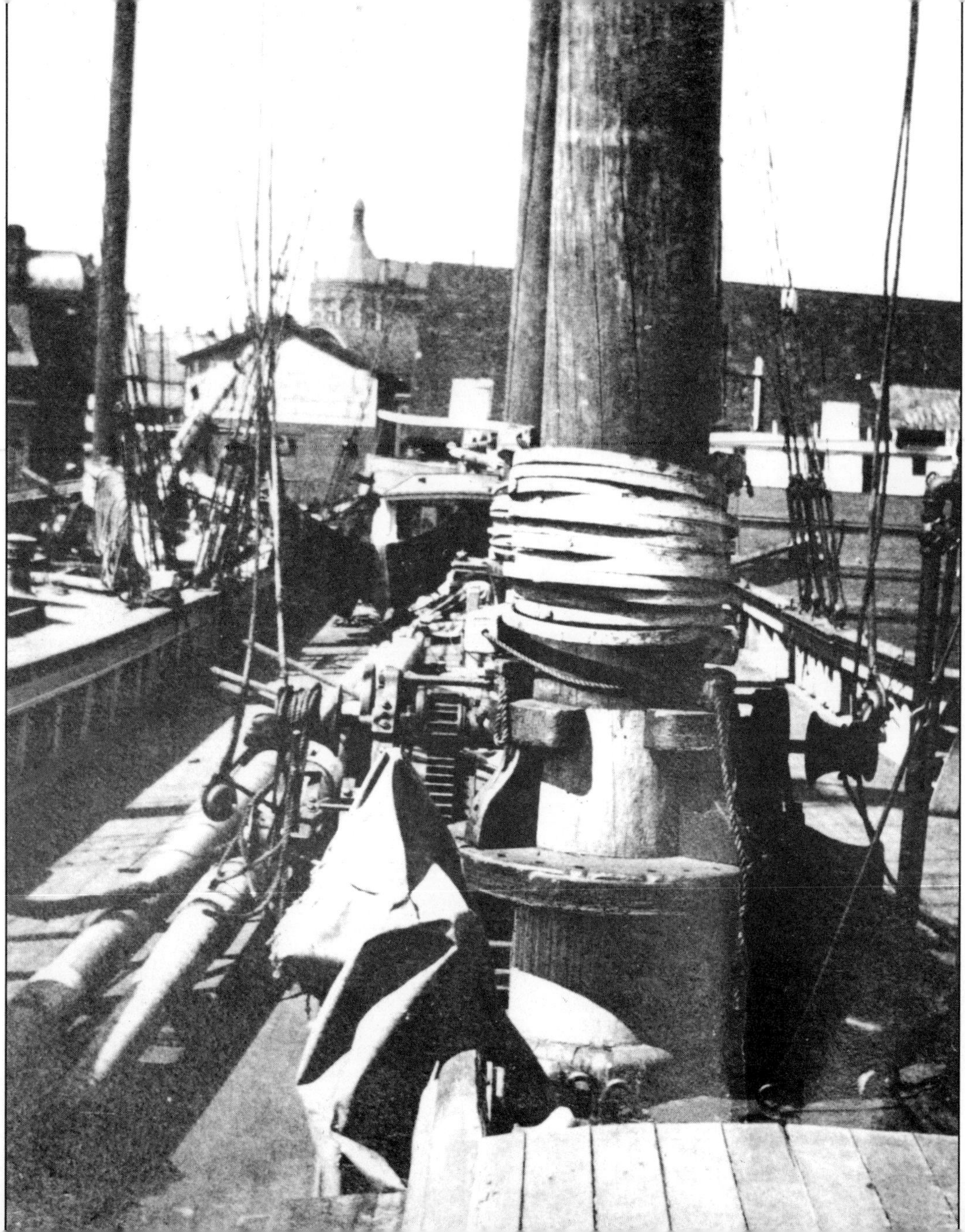

Donkey Engine of the Emerald

On the foredeck of the *Emerald*, visible at the right of the picture can be seen the boiler of the steam donkey engine used to assist the crew in hauling the lines, weighing anchor and handling cargo. The winch powered by the donkey is aft of the mast. The steam donkey allowed schooners to compete with larger powered ships for much longer than might otherwise have been the case by increasing the effectiveness of their relatively small crews (see also pg. 51).

Marine Museum of Upper Canada, Toronto

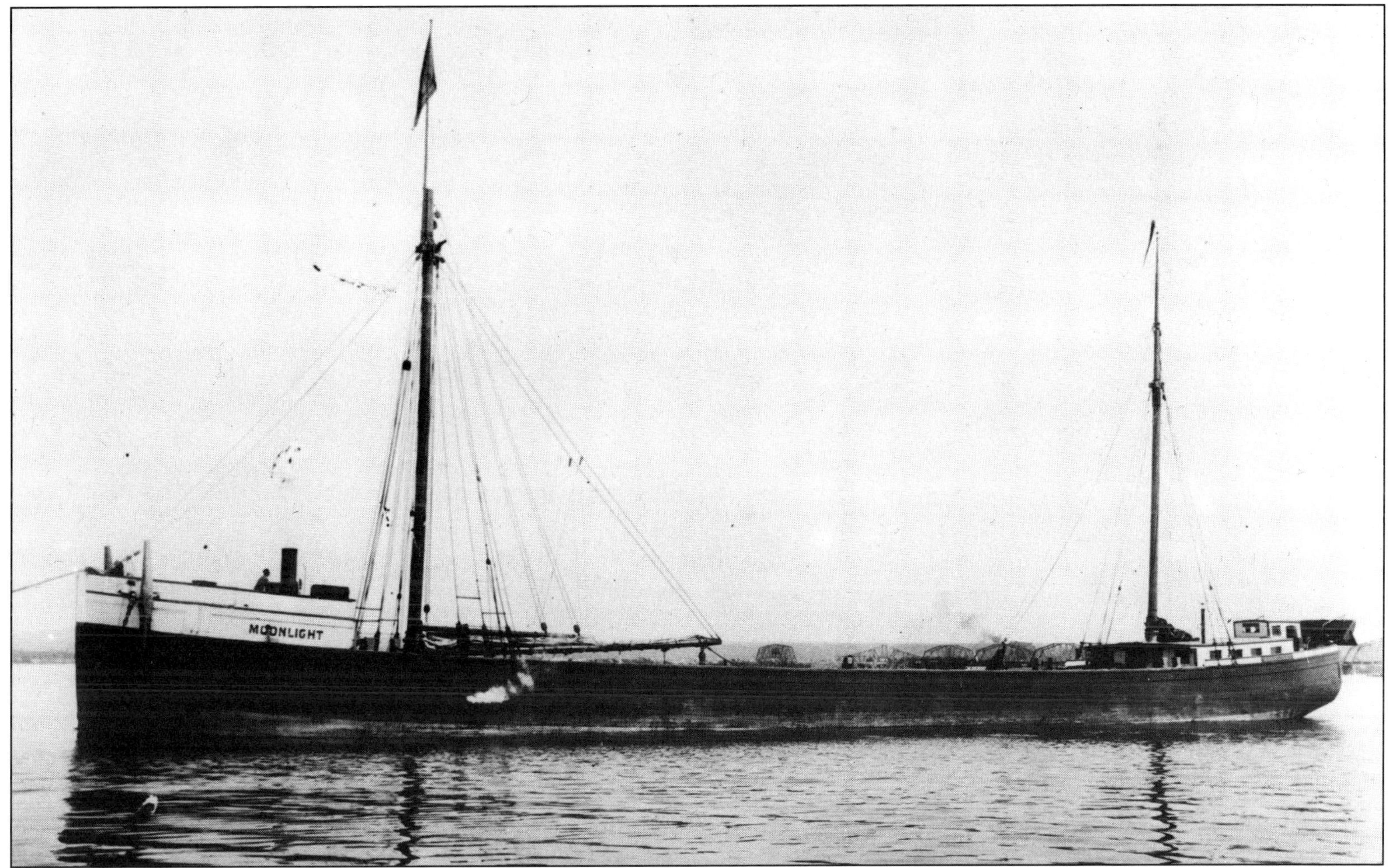

Great Lakes Historical Society, Vermilion, Ohio

Moonlight

The *Moonlight* was built in Milwaukee, Wisconsin, in 1874 and was considered one of the stars of the day. It had once raced from Buffalo home to Milwaukee against the schooner *Porter*. *Moonlight* lost when a gale blew up as the two entered their destination port. It ended its days in the configuration shown here—as a tow barge. Sometimes these barges were towed by tugs and sometimes they were towed as consorts behind cargo steamers. Enough canvas was left to enable these semi-retired schooners to manoeuvre a little if they were cut loose from the vessels towing them. On occasion they were able to survive, but in 1903 *Moonlight* had no such luck when it parted from the steamer *Charles J. Kershaw* and went aground in Lake Superior.

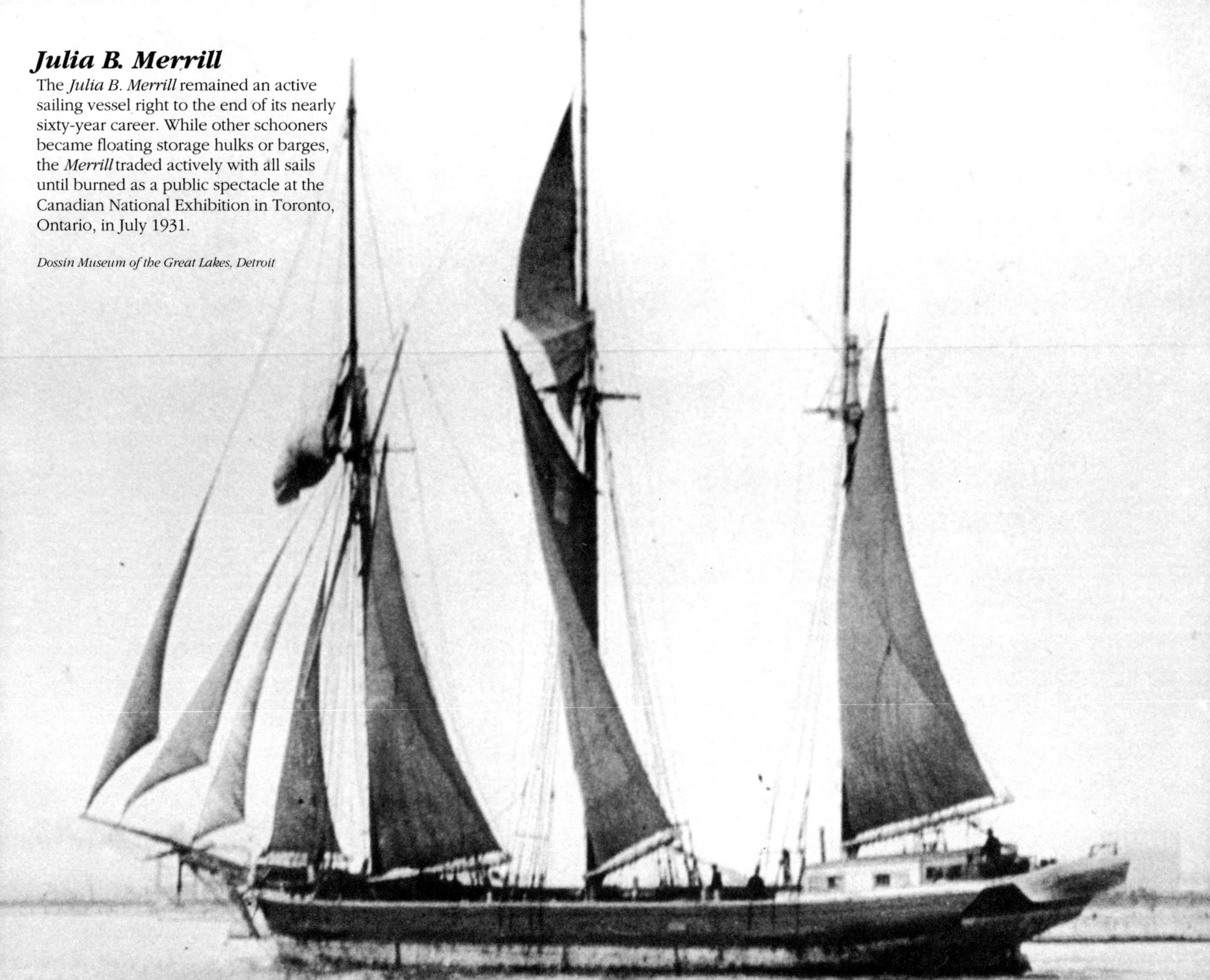

Julia B. Merrill

The *Julia B. Merrill* remained an active sailing vessel right to the end of its nearly sixty-year career. While other schooners became floating storage hulks or barges, the *Merrill* traded actively with all sails until burned as a public spectacle at the Canadian National Exhibition in Toronto, Ontario, in July 1931.

Dossin Museum of the Great Lakes, Detroit

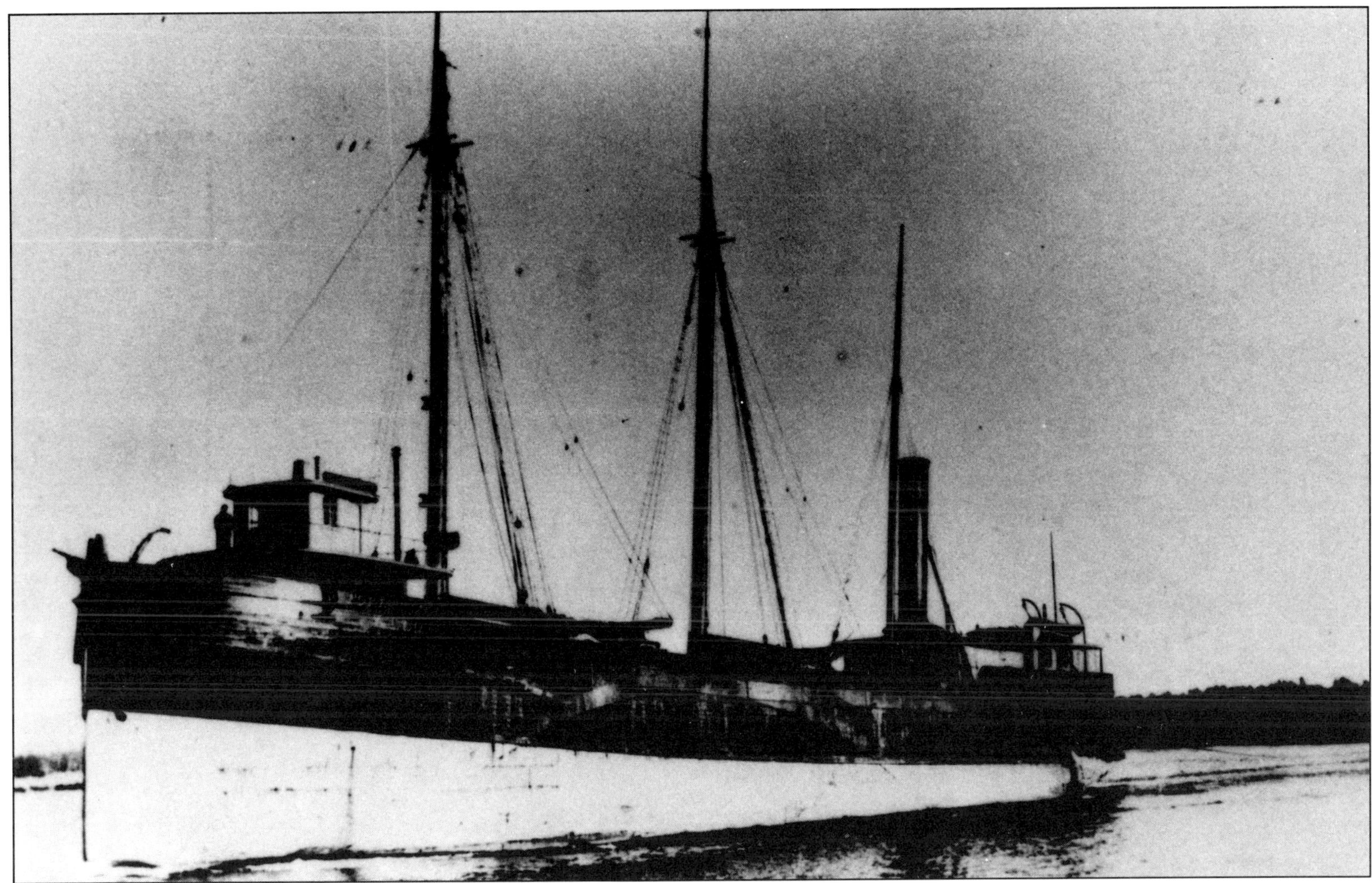

Buffalo and Erie County Historical Society

Forest City

The *Forest City* was built in Cleveland, Ohio, in 1870, very much in the height of the days of sail. But in this picture, taken sometime before being wrecked in Georgian Bay in 1904, it has been completely transformed to a steamer with many of the characteristics that dominated laker design for generations. We can see the forward wheelhouse, the funnel rising through the after cabin and the proportions that enabled canal transiting. Only the masts are a link to its former life.

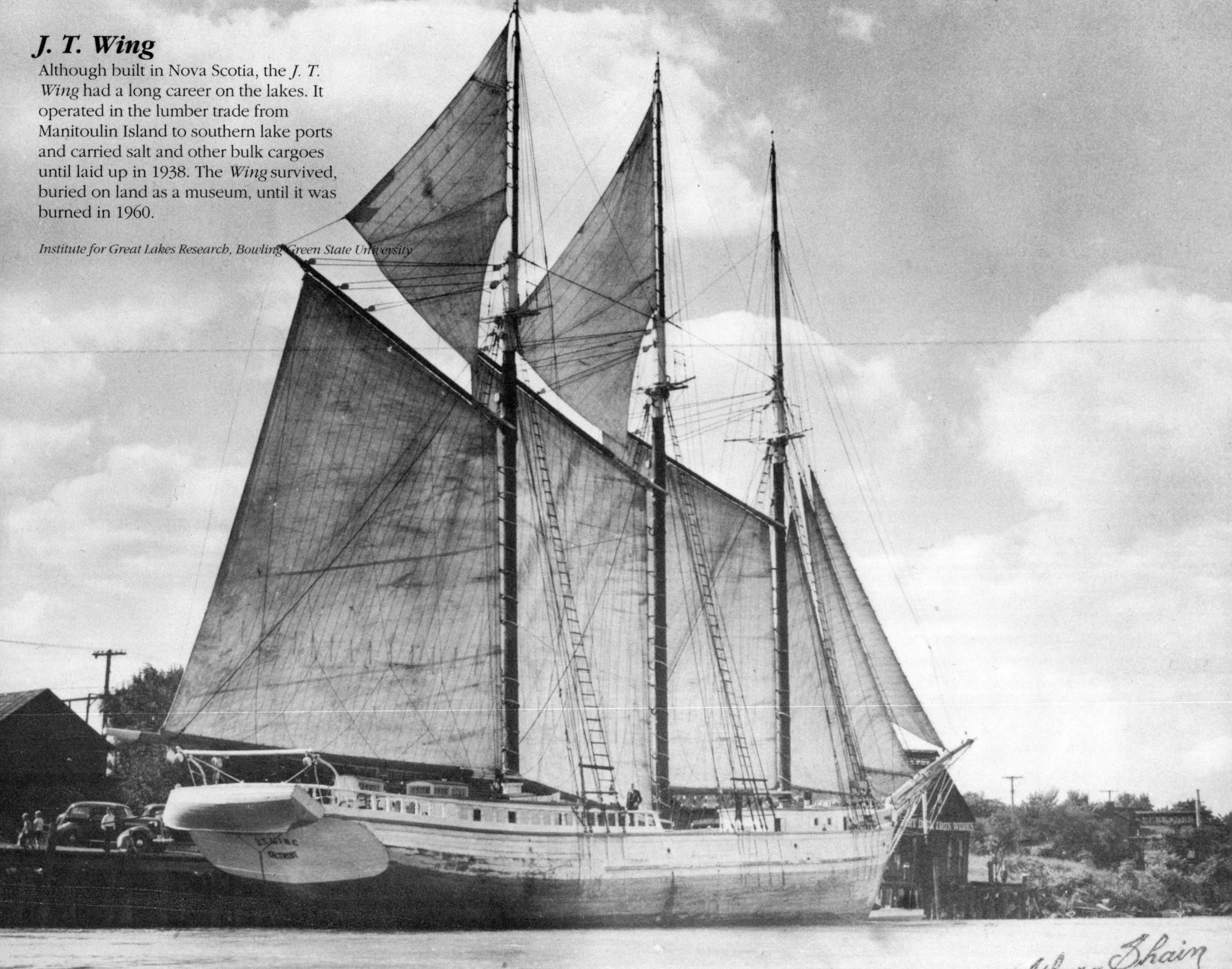

J. T. Wing

Although built in Nova Scotia, the *J. T. Wing* had a long career on the lakes. It operated in the lumber trade from Manitoulin Island to southern lake ports and carried salt and other bulk cargoes until laid up in 1938. The *Wing* survived, buried on land as a museum, until it was burned in 1960.

Institute for Great Lakes Research, Bowling Green State University

FURTHER READING

Addis, Fred and John N. Jackson. *The Welland Canals: A Comprehensive Guide.* St. Catharines, Ontario: Lincoln Graphics, 1982.

Greenhill, Basil. *Schooners.* London: B. T. Batsford Ltd., 1980.

Greenhill, Basil. *The Merchant Schooners.* London: B. T. Batsford Ltd., 1983.

Greenwood, John O. *Namesakes 1930-1955.* Cleveland, Ohio: Freshwater Press Inc., 1978.

Greenwood, John O. *Namesakes 1920-1929.* Cleveland, Ohio: Freshwater Press Inc., 1984.

Greenwood, John O. *Namesakes 1910-1919.* Cleveland, Ohio: Freshwater Press Inc., 1986.

Greenwood, John O. *Namesakes 1900-1909.* Cleveland, Ohio: Freshwater Press Inc., 1987.

Kuttruff, Karl. *Ships of the Great Lakes, A Pictorial History.* Detroit, Michigan: Wayne State University Press, 1976.

Metcalfe, Willis. *Marine Memories.* Picton, Ontario: The Picton Gazette Publishing Co.(1971) Limited, 1975.

Shipley, Robert J. *St. Catharines, Garden on the Canal.* Burlington, Ontario: Windsor Publications, 1987.

Wilson, Garth S. *Great Lakes Historic Ships Research Project, The Documentation and Analysis of the 19th Century Great Lakes Wooden Hull Design.* Kingston, Ontario: Marine Museum of the Great Lakes, 1989.

ACKNOWLEDGEMENTS

As well as thanking their families for constant support, especially Pamela Shipley Fielding for her hours of research and proofreading, the authors would like to acknowledge the encouragement of their employers, The Harbourfront Corporation of Toronto and The Welland Canals Society. Thanks also to the following collections and collectors of pictures and historical information: The Buffalo and Erie County Historical Society; Mr. John Burtniak of Brock University Library; The Dossin Museum of the Great Lakes in Detroit; The Great Lakes Historical Society of Vermilion, Ohio; The Institute for Great Lakes Research at Bowling Green State University; The Manitowoc Maritime Museum; The Marine Museum of the Great Lakes at Kingston; The Marine Museum of Upper Canada in Toronto; The National Archives of Canada; The Oakville Historical Museum; The Public Archives of Ontario; Mr. Al Sagon-King of Thorold; and The St. Catharines Historical Museum.